T0135923

Suprathreshold perception in normal-hearing and hearing-impaired listeners

Dissertation

zur Erlangung des akademischen Grades

**doctor rerum naturalium
(Dr. rer. nat.)**

genehmigt durch die Fakultät für
Naturwissenschaften
der Otto-von-Guericke-Universität Magdeburg

von: Dipl.-Phys. Jan Hots

geb. am   12. Oktober 1983 in Oldenburg

Gutachter: Prof. Dr. rer. nat. Jesko L. Verhey

                     Prof. Dr. rer. nat. Torsten Dau

eingereicht am: 13. Dezember 2013

verteidigt am: 4. Juli 2014

Bibliografische Information der Deutschen Nationalbibliothek

Die Deutsche Nationalbibliothek verzeichnet diese Publikation in der
Deutschen Nationalbibliografie; detaillierte bibliografische Daten sind
im Internet über http://dnb.d-nb.de abrufbar.

©Copyright Logos Verlag Berlin GmbH 2014
Alle Rechte vorbehalten.

ISBN 978-3-8325-3758-6

Logos Verlag Berlin GmbH
Comeniushof, Gubener Str. 47,
10243 Berlin
Tel.: +49 (0)30 42 85 10 90
Fax: +49 (0)30 42 85 10 92
INTERNET: http://www.logos-verlag.de

Kurzfassung

Die vorliegende Arbeit befasst sich mit der überschwelligen Wahrnehmung von Geräuschen im normalen und im pathologischen Gehör. Trotz intensiver Erforschung des Gehörs über viele Jahrzehnte, ist das menschliche Hören bislang nicht vollständig verstanden. Für die Entwicklung vieler Anwendungen, die unser alltägliches Leben verbessern können (z.B. die Abschätzung von Lärmemissionen oder die Entwicklung von Hörgeräten), ist ein besseres Verständnis des komplexen Prozesses der Wahrnehmung von Geräuschen allerdings von großer Bedeutung. Unterschiedliche Bereiche der aktuellen Forschung befassen sich derzeit mit der Funktionsweise des Gehörs. Der Fokus dieser Arbeit liegt auf der psychoakustischen Beschreibung der Wahrnehmung von Schallintensität und Lautheit sowie deren Beeinflussung durch Hörstörungen. Neben der Intensität eines Geräusches wird dessen Lautheit durch verschiedene Parameter beeinflusst. In dieser Arbeit werden Lautheitseffekte untersucht, die mit der zeitlichen Struktur und dem spektralen Gehalt eines Geräusches zusammenhängen.

Mit zunehmender Dauer eines Signals steigt auch die wahrgenommene Lautheit an. Dieser Effekt wird im Allgemeinen als zeitliche Lautheitssummation bezeichnet und mit Hilfe eines Tiefpasses simuliert, dessen Zeitkonstante wenige Millisekunden beträgt. In der vorliegenden Arbeit wird gezeigt, dass eine einfache Erweiterung dieses Modellansatzes durch eine Kompressivität, wie sie im auditorischen System zu finden ist, zu deutlichen Abweichungen der Vorhersagen von Literaturdaten der zeitlichen Lautheitssummation führt. Dieses Problem kann durch einen neuen Modellansatz, in dem zwei Tiefpässe mit unterschiedlichen Zeitkonstanten in paralleler Weise angeordnet sind, gelöst werden. Eine Integration dieses Ansatzes in die zeitliche Stufe aktueller dynamischer Lautheitsmodelle führt zu Verbesserungen der vorhergesagten Daten.

Für den Einfluss des spektralen Gehaltes auf die Lautheit werden in der klassischen Vorstellung zwei Fälle unterschieden. Ist die Bandbreite des Geräusches größer als eine kritische Bandbreite, so steigt bei gleichbleibender Intensität die Lautheit mit zunehmender Bandbreite an. Dieser Effekt wird als spektrale Lautheitssummation bezeichnet. Für den Fall, dass die Bandbreite kleiner ist als die kritische Bandbreite gehen stationäre Lautheitsmodelle davon aus, dass die Bandbreite keinen Einfluss auf die Lautheit hat. In der vorliegenden Arbeit wird gezeigt, dass aktuelle dynamische Lautheitsmodelle hingegen eine negative Pegeldifferenz zwischen subkritischen Rauschsignalen und einem Sinuston bei gleicher Lautheit vorhersagen. Im Experiment mit normalhörenden Versuchspersonen wird für verschiedene Mittenfrequenzen und Referenzpegel ein gegenteiliger Effekt gefunden, d.h. eine positive Pegeldifferenz bei gleicher Lautheit. Dieser Effekt ist für Bandbreiten nahe der kritischen Bandbreite am größten und nimmt mit steigendem Referenzpegel ab. Gleiche Ergebnisse konnten mit unterschiedlichen Referenzsignalen und Messverfahren erzielt werden, was davon zeugt, dass dieser Effekt eine grundlegende Eigenschaft des Gehörs ist. Um dem Ursprung des zugrunde liegenden Vorgangs im Gehör näher zu kommen, wurden vergleichbare Experimente mit schwerhörenden Versuchspersonen durchgeführt. Für diese ist der gefundene Effekt, verglichen mit den Normalhörenden, reduziert und das Maximum zu größeren Bandbreiten verschoben. Dieses Ergebnis lässt auf eine Ursache auf cochleärer Ebene schließen.

Die Anhebung der Hörschwelle wurde für lange Zeit als Indikator einer Hörstörung angesehen. Im Gegenzug wurde bei normalen Hörschwellen auch von einem intakten Gehör ausgegangen. Aktuelle Studien zeigen jedoch, dass Lärmexposition bei Pegeln, die zu einer vollständig reversiblen Erhöhung der Hörschwelle führen mit einer dauerhaften Deafferenzierung auditorischer Nervenfasern einhergehen, die zu einem Tinnitus führen können. In der vorliegenden Arbeit wurden gerade unterscheidbare Änderungen der Intensität bei normalhörenden Versuchspersonen und Versuchspersonen mit Tinnitus und einer normalen Hörschwelle untersucht. Die Ergebnisse dieser Messung zeigen eine verringerte Auflösung der Intensität bei mittleren Pegeln im Frequenzbereich des Tinnitus. Dies ist ein Hinweis auf eine Deafferenzierung der auditorischen Nervenzellen und bestätigt diese als mögliche Ursache für Tinnitus.

Abstract

Loudness perception, among other aspects, depends on temporal and spectral properties of a sound. In the present thesis temporal and spectral effects of loudness as well as the changing of the perception of loudness and sound intensity in hearing-impaired listeners are investigated.

Loudness increases with duration, an effect commonly referred to as temporal integration of loudness and characterized by time constants of leaky integrators. It is shown that such a model approach is unable to predict the slope of the temporal integration function for loudness when the compressive characteristics of the auditory system are taken into account. Deviations are still observed when a more sophisticated loudness model is used. A good prediction is achieved when the single low-pass filter stage of the loudness model is replaced with a stage consisting of two parallel low-pass filters with different time constants.

Loudness also increases with bandwidth, as soon as a critical bandwidth is exceeded, an effect commonly referred to as spectral loudness summation. For sounds with a subcritical bandwidth, stationary loudness models assume that loudness is determined by the intensity and center frequency only. This is not in agreement with dynamic loudness models, which predict higher levels for a tone than for equally loud noises. Experimental data of normal-hearing listeners for different center frequencies, noise bandwidths and reference levels show the opposite effect, i.e., a positive level difference between noise and pure tone at equal loudness. The results indicate that the decrease of loudness at equal level with increasing subcritical bandwidth is a common property of the auditory system. To gain more information about the origin of this effect, normal-hearing data are compared to results of a similar experiment with hearing-impaired participants. The latter show a reduced effect and a shift of the effect's maximum towards broader bandwidths.

Recent measurements in tinnitus patients with normal audiograms indicate the presence of hidden hearing loss due to reduced neural output from the cochlea at suprathreshold levels. Results of this thesis show that intensity discrimination thresholds were significantly increased in the tinnitus frequency range of these listeners, consistent with the hypothesis that auditory nerve fibre deafferentation is associated with tinnitus.

Contents

Contents

1 Introduction

1.1 Motivation

In our everyday life sounds surround us almost constantly. They provide an important tool for our communication, they help us to orientate, e.g., in traffic, they can enrich our lives in the form of music, and unfortunately they can also be very annoying to us in the form of noise emissions. The perception of sound by the human auditory system is a complex process, which cannot be understood by the physical description of sound waves alone. When a sound wave reaches our ear, the tympanic membrane and the auditory ossicles transform it to a pressure wave in the fluid of our cochlea. The hair cells on the basilar membrane process this pressure wave to neuronal information, which is conducted to our brain by the auditory nerve, where finally the percept of the sound is generated. Despite the intensive investigation of the auditory system for many decades, the perception of sound is not completely understood, yet. However, for many applications in our everyday life, be it the estimation of noise emissions, the design of effective but not annoying warning sounds in electromobility or the development of hearing aids, a better understanding of this complex process is of major importance. One field of research that investigates the mechanics of the human auditory system and the perception of sound is psychoacoustics. Different psychoacoustic measures such as the roughness, the impulsiveness, the magnitude of tonal content, the pitch strength or the loudness of a sound help to characterize its perception. The scope of this thesis is the perception of loudness and sound intensity.

1.2 Loudness perception

"Loudness is the primary perceptual correlate of the level of a sound" (Florentine, 2011). In the infancy of psychophysics it was assumed that the internal representation of intensity could be assessed by just noticeable differences in the intensity. It was shown that the relative change in level $\frac{\Delta I}{I}$ is roughly constant over a large level range (see, e.g., Viemeister and Bacon, 1988). This is commonly referred to as Weber's law. However, later studies argued that loudness as the sensation associated with intensity cannot be assessed by just noticeable differences. Instead, it has to be rated by other means, such as magnitude estimation, which leads to the nowadays common Sone scale (see, e.g., Fastl and Zwicker, 2007). Loudness also changes with other physical stimulus properties (see, e.g., Moore, 2003; Fastl and Zwicker, 2007; Florentine, 2011). The influence of the level as well as the influence of spectral and temporal characteristics of a sound on loudness will be introduced in the following.

1.2.1 Level effects

The loudness N of a sound is highly correlated with the intensity of the sound. A sound has the loudness of 1 Sone if it is perceived equally loud as a 1-kHz pure tone at a level of 40 dB SPL (compare Moore, 2003; Fastl and Zwicker, 2007). For intermediate to high levels a level increase of about 10 dB leads to a doubling of the loudness in Sone. At low levels a lower increase of the level suffices to double the perceived loudness. This level dependency of the loudness of a sound is commonly visualized by loudness functions, showing the loudness in Sone on a logarithmic scale as a function of the level in dB SPL. For inter-

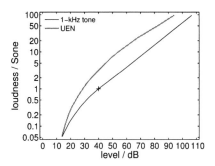

Figure 1.1: Loudness functions of a 1-kHz pure tone (black line)
and *uniform exciting noise* (UEN, gray line). The cross
indicates the loudness of 1 Sone of the 1-kHz pure tone
at a level of 40 dB SPL. (Data replotted from Fastl and
Zwicker, 2007, Fig. 8.4)

mediate levels a linear relation between loudness and level with
a slope smaller than one is found in this depiction. This slope is
usually interpreted as a compressive exponent in the perception
of loudness. Figure 1.1 shows the loudness functions of a 1-kHz
pure tone (black line) and *uniform exciting noise* (gray line,
see Fastl and Zwicker, 2007). The data is replotted from Fastl
and Zwicker (2007) (see their Fig. 8.4). The cross in the figure indicates the loudness of the 1-kHz pure tone at a level of
40 dB SPL, which amounts to 1 Sone.

1.2.2 Spectral effects

The loudness evoked by a sound also depends on the bandwidth of the sound. The loudness functions in Fig. 1.1 show,
that the loudness of the noise signal is always higher than that
of the pure tone at the same level. Zwicker *et al.* (1957) showed

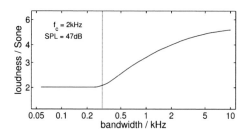

Figure 1.2: Loudness of a bandpass noise centered at 2 kHz with a constant level of 47 dB SPL as a function of the noise width. The vertical gray line indicates the critical bandwidth. (Data replotted from Fastl and Zwicker, 2007, Fig. 6.7)

that at constant level the loudness of a stimulus is independent of the stimulus bandwidth up to a certain bandwidth denoted as the critical bandwidth. If the stimulus bandwidth is increased beyond the critical bandwidth, the loudness increases with bandwidth. Figure 1.2 shows this effect for a noise signal centered at 2 kHz at a constant level of 47 dB SPL. The loudness of the noise is plotted as a function of the noise width. The data is replotted from Fastl and Zwicker (2007) (see their Fig. 6.7). This effect is commonly referred to as spectral loudness summation. It can be explained by the assumption that the auditory system firstly generates the loudness in critical band filters and then integrates across the critical band filters. The loudness generation in each of the critical bands is compressive respective to the intensity of the stimulus. If the entire energy of a stimulus falls into a single critical band filter, the resulting loudness is lower than in the case the energy is spread over several critical band filters. Thus, the size of the spectral loudness

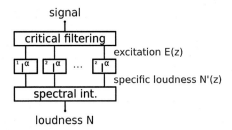

Figure 1.3: General structure of a stationary loudness model. In
the first stage the long-term spectrum of the input signal is divided into z critical band filters resulting in
the excitation E(z). In the following compressive stage
intensity values are raised to the power of α resulting
in the specific loudness N'(z). An integration across
the critical band filters leads to the loudness N of the
signal.

summation is determined by the power of the compression and
the width of the critical band filters.

To model the effect of spectral loudness summation, the specific loudness N'(z) of a sound has to be determined from the
excitation E(z) by a power law in every critical band filter z.
A summation across all critical band filters results in the loudness N of the sound. The general structure of such a stationary
model approach is schematically shown in Fig. 1.3. In the first
stage the signal (usually the long-term spectrum of the sound)
is divided into critical band filters, resulting in the excitation
E of every critical band filter z. In the next stage the specific
loudness N'(z) is determined from the excitation. Specific loudness and excitation are related by $N'z \sim E(z)^{\alpha}$, where $\alpha < 1$ is
the compression factor. In the last stage of the model the spe-

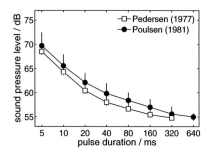

Figure 1.4: Temporal integration data of Pedersen *et al.* (1977)
(open squares) and Poulsen (1981) (filled circles, error
bars indicate 95% confidence limits) for a reference level
of 55 dB SPL. The sound pressure level of the matching
tone pulse is shown as a function of the pulse duration.
(Data replotted from Poulsen, 1981, Fig. 1)

cific loudness $N'(z)$ is integrated across all critical band filters
z, resulting in the loudness N of the sound.

Different stationary loudness models accounting for the spectral effects of loudness were already introduced (see, e.g., Fletcher
and Munson, 1937; Zwicker, 1958; Zwicker and Scharf, 1965;
Moore *et al.*, 1997; Moore and Glasberg, 2007). These models
work on long-term spectra of the sound and do not regard its
temporal structure.

1.2.3 Temporal effects

Apart from the spectral content of a sound, loudness also
depends on its duration and temporal structure. In general,
longer stimuli evoke a higher loudness than short stimuli. This
effect is referred to as temporal integration of loudness. It

can be explained by the sluggishness of the auditory system. Figure 1.4 shows experimental results of Pedersen *et al.* (1977) (open squares) and Poulsen (1981) (filled circles). In both studies 1-kHz test tone pulses of different durations from 1 to 320 ms (Pedersen *et al.*, 1977) or from 1 to 640 ms (Poulsen, 1981) were matched in loudness to a 1-kHz reference tone pulse with a level of 55 dB SPL and a duration of either 320 ms (Pedersen *et al.*, 1977) or 640 ms (Poulsen, 1981). In Fig. 1.4 the level of the test tone pulse at equal loudness is shown as a function of duration. Error bars in the data of Poulsen (1981) indicate 95% confidence limits. In both studies the level for the short tone pulses has to be markedly raised to evoke an equal loudness. To predict these data, Poulsen (1981) suggests a simple leaky integrator model with a time constant of $\tau = 80$ ms for the data by Pedersen *et al.* (1977) or of $\tau = 110$ ms for the other data set. Alternatively, he suggests two consecutive leaky integrators, one with a short and one with a long time constant of $\tau_1 = 10$ ms and $\tau_2 = 50$ ms or $\tau_1 = 4$ ms and $\tau_2 = 100$ ms concerning the two sets of data.

The combination of such a leaky integrator model and the stationary model described in Sec. 1.2.2 leads to a dynamic loudness model, that processes the spectral as well as the temporal information of a sound. The general structure of such a model is shown in Fig. 1.5. In the first model stage the time signal of the sound is sampled by a temporal window with a duration of a few milliseconds. In the spectral stage every resulting time segment is processed similar to the processing of a stationary model, i.e., this stage consists of a critical band filtering, a compression and a spectral integration. At the end of this stage the instantaneous loudness for each time segment $N_{inst}(t)$ is received. In the temporal integration stage this instantaneous loudness is low-pass filtered, resulting in the loudness of each time segment $N(t)$. In the last stage the overall loudness N of the sound is determined. Usually, the maximum or a percentile of the time dependent loudness is determined in this stage.

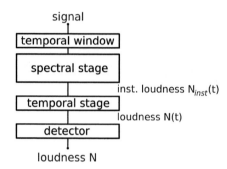

Figure 1.5: General structure of a dynamic loudness model. In the spectral stage the signal is processed similar to the processing in a stationary model. In contrast to the stationary model the temporal structure of the sound is considered by the temporal window. The resulting instantaneous loudness $N_{inst}(t)$ is low-pass filtered in the temporal stage leading to the time dependent loudness $N(t)$. In the final stage the loudness N of the sound is detected from the time dependent loudness $N(t)$.

Different dynamic loudness models of this character have been presented like, e.g., the procedure according to DIN 45631/A1 (2010), the model of loudness applicable to time-varying sounds (*TVL*) by Glasberg and Moore (2002) or the dynamic loudness model (*DLM*) by Chalupper and Fastl (2002), and an extended version of this model (*eDLM*) by Rennies *et al.* (2009).

The perception of loudness is also influenced by amplitude modulations of a sound (see, e.g., Moore *et al.*, 1999; Grimm *et al.*, 2002; Fastl and Zwicker, 2007). Stimuli with slow amplitude modulations are perceived louder than unmodulated stimuli. The size of this effect varies between the studies and corresponds to a level difference from -1 dB (compare Grimm *et al.*, 2002) to -3 dB (compare Moore *et al.*, 1999; Fastl and Zwicker, 2007) between the amplitude modulated and the unmodulated signal at equal loudness. This effect can be predicted by a maximum detector in a dynamic loudness model. Grimm *et al.* (2002) showed that this effect is larger for stimuli with a broad bandwidth than for narrower stimuli. In contrast to these findings Grimm *et al.* (2002) found that the loudness of an amplitude modulated pure tone is lower than that of an unmodulated pure tone.

1.2.4 Spectro-temporal effects

Besides the spectral and temporal effects described above, there are spectro-temporal effects, that affect the perception of loudness. Although these effects are not directly in the scope of this thesis, two of them shall be described shortly, since the results of this thesis may contribute to their understanding.

The spectral content of frequency modulated sounds varies with time. Thus, in the determination of the loudness of these sounds the spectral as well as the temporal content need to be considered. Zwicker (1974) compared the loudness of frequency modulated and unmodulated 1.5-kHz pure tones with a modulation depth of 700 Hz and modulation frequencies from 1 to 512 Hz to the loudness of a noise signal centered at 1.5 kHz with a level of 50 dB SPL. For the high modulation frequencies the study found a level difference of about 5 dB between the unmodulated and the frequency modulated pure tone at equal loudness, whereas for the low modulation frequencies this effect decreased. One explanation of this effect is the ability of the auditory system to follow the change of the spectral content with time at low modulation frequencies. For higher modulation frequencies the temporal resolution of the auditory system is not sufficient to follow the modulation, with the result that effects of spectral loudness summation occur.

The studies of Verhey and Kollmeier (2002), Anweiler and Verhey (2006), and Verhey and Uhlemann (2008) showed that the amount of spectral loudness summation depends on the duration of the stimuli. It was largest at short durations of 10 ms and decreased with an increase of duration for signals of 100 and 10 ms. The mechanism underlying this effect is still unclear. Rennies *et al.* (2009) excluded an adaption of compression for different stimulus durations as hypothesized by Verhey and Kollmeier (2002). It was shown that the effect on the predicted spectral loudness summation was rather small, while the slopes of the loudness functions changed for the different signals. This was not found in experimental data. Verhey and Uhlemann (2008) alternatively suggested adaptive auditory filters or the influence of higher cognitive processes as the underlying mechanism. Rennies *et al.* (2009) showed that by a bandwidth dependent onset enhancement in the *DLM* their model can account for the effect between durations of 10 and 1000 ms. However, at intermediate durations their model still cannot predict the difference in spectral loudness summation.

1.3 The influence of hearing-impairment on our perception

Our auditory processing is a vulnerable system and can easily get harmed (e.g., by sounds with a high intensity or by diseases). An impairment of this system changes the perception of sound. Roughly, it can be distinguished between two cases of cochlear damage (see, e.g., Launer, 1995; Moore and Glasberg, 1997; Moore, 2003). If the hearing loss is based on a loss of the inner hair cells in our cochlea, the hearing threshold is elevated and the level has to be increased by a constant value for the whole level range to evoke an equally loud perception of the signal compared to a normal-hearing listener. Thus, the loudness function is horizontally shifted towards higher levels compared to the loudness function of a normal-hearing listener. If, in contrast, the hearing loss is based on a loss of the outer hair cells in our cochlea, the hearing threshold is elevated as well and a higher level is needed for a signal to be detected. For high levels, however, the loudness function converges to the loudness function of a normal-hearing listener. Thus, the dynamic range in the auditory system is reduced and the compressive relation between level and loudness is changed. This is referred to as recruitment. It can be explained by the amplifying characteristics of the outer hair cells (see, e.g., Moore, 2003).

The change in the slope of the loudness function, which is commonly associated with a reduced compression in hearing-impaired listeners, leads to further changes in the perception of sound. Moore *et al.* (1996) showed that the modulation depth of amplitude modulated sounds is higher in the impaired ear. Since model approaches as described in Sec. 1.2.3 use the maximum or a percentile of the time dependent loudness as

11

an estimate of the overall loudness of a sound, the loudness of amplitude modulated sounds changes with a change in the perception of the modulation depth. Different studies also found a broadening of the auditory filters in hearing-impaired listeners compared to normal-hearing listeners (see, e.g., Nitschmann *et al.*, 2010; Baker and Rosen, 2002). Thus, since spectral loudness summation is assumed to be due to critical band filtering and compression, it should be reduced in listeners with an impaired hearing. This was approved by the studies of Verhey *et al.* (2006) and Brand and Hohmann (2001).

The impairment of the auditory system as described above involves an elevation of the hearing threshold in the case of the loss of inner, as well as outer hair cells. For long, normal hearing thresholds have been regarded as an indicator of an intact auditory system. However, in a physiological study Kujawa and Liberman (2009) showed that the exposure to noise at a level leading to a temporal shift of the threshold caused a permanent deafferentation of auditory nerve (AN) fibres in the high frequency range. This cochlear damage was approved by a reduction of the wave I amplitude in auditory brainstem responses (see Kujawa and Liberman, 2009). Their results suggested that cochlear damage induced by noise has consequences to our hearing that cannot be revealed by a conventional threshold testing, but may contribute to tinnitus.

1.4 Structure of the thesis

As described in Sec. 1.2.2 the loudness of stationary sounds, i.e., sounds with a constant level and spectral content over time is widely understood and can be predicted by stationary loudness models. Most of the sounds in our environment (e.g., speech), however, are not of a stationary nature, but show a dynamic structure, i.e., their level and spectral content vary with time. To predict the loudness of these sounds, dynamic loudness models as described in Sec. 1.2.3 are used. Rennies *et al.* (2010) showed that some experimental results are still challenging to these types of models. They showed, that at intermediate durations the predictions of the *DLM* as well as the *TVL* deviated from the experimental results on temporal integration of loudness of Pedersen *et al.* (1977) and Poulsen (1981) shown in Fig. 1.4. In Chapter 2 of this thesis it is shown that similar deviations occur in the leaky integrator models proposed by Poulsen (1981), as soon as nonlinear characteristics of the auditory system are included. It is tested, to what extent the predictions by a model approach using two leaky integrators arranged in a parallel fashion match with the experimental data at all durations. Using the example of the *eDLM* it is studied to what extent the predictions of dynamic loudness models for intermediate durations can be improved if this model approach is included in the temporal integration stage.

The approach to model spectral effects of loudness described in Sec. 1.2.2 predicts a loudness not depending on bandwidth as long as the critical bandwidth is not exceeded. Scharf (1978) pointed out that the loudness of noise signals with a very narrow bandwidth may be affected by inherent level fluctuations. In Chapter 3 the predictions of dynamic loudness models for noise bands with a subcritical bandwidth (i.e., a bandwidth narrower than the critical bandwidth) are compared to results of a loudness matching experiment. Chapter 4 extends these data by results for other center frequencies, reference levels and measurement procedures. The experimental results of both chapters show a dependency of loudness on bandwidth, that is not compatible with the common model approaches of loudness. The underlying mechanism of the found effect is still unclear. To test the hypotheses about the underlying mechanism made in these chapters, similar experiments with hearing-impaired listeners were conducted, which are described in Chapter 5.

As described in Sec. 1.3 Kujawa and Liberman (2009) suggested that the deafferentation of AN fibres may contribute to tinnitus. A link between deafferentation of AN fibres and the development of tinnitus is also indicated by computational modeling (see Schaette and McAlpine, 2011). Since in our auditory system the resolution of the sound intensity is coded by AN fibres (Liberman and Kiang, 1978; Yates *et al.*, 1990), deafferentation should reduce the ability to discriminate differences in the sound intensity. Thus, if participants reporting a tinnitus have a deafferentation of AN fibres, their performance in an intensity discrimination task should be reduced compared to the normal-hearing listeners. To test this hypothesis, just-noticeable intensity differences were measured in normal-hearing listeners and listeners with a normal audiogram who reported a tinnitus in Chapter 6.

A modified Version of Chapter 2 is published in "Acta Acustica united with Acustica" as:

Hots, J., Rennies, J. and Verhey, J.L. **(2014)** "Modeling temporal integration of loudness" Acta Acust. united Ac. **100** (1) 184–187.

Modified versions of the Chapters 3, 4 and 6 are published in the "Journal of the Acoustical Society of America" as:

Hots, J., Rennies, J. and Verhey, J.L. **(2013)** "Loudness of sounds with a subcritical bandwidth: a challenge to current loudness models?" J. Acoust. Soc. Am. **134** (4) EL334–EL339.

Hots, J., Rennies, J. and Verhey, J.L. "Loudness of subcritical sounds as a function of bandwidth, center frequency, and level" J. Acoust. Soc. Am. **135** (3) 1313–1320.

Epp, B., Hots, J., Verhey, J.L. and Schaette, R. **(2012)** "Increased intensity discrimination thresholds in patients suffering from tinnitus with a normal audiogram" J. Acoust. Soc. Am. **132** (3) EL196–EL201.

In Epp *et al.* (2012) (Chapter 6) the main contributions of the autor of this thesis were the conduction of the experiments and part of the data analysis. In all the other studies, the design, conduction and analysis of the studies were done by the author.

2 Modeling temporal integration of loudness[1]

2.1 Introduction

The fundamental psychoacoustical sensation loudness is close-ly related to the physical intensity of the sound but also depends on other stimulus properties such as spectrum and temporal characteristics. Loudness is an important factor when assessing noise emissions. As a consequence, loudness was standardized in several national and international standards (DIN 45631/A1, 2010; ANSI S3.4, 2007; ISO 226, 2003; ISO 532, 1975). Since many environmental sounds vary in their intensity over time it is important to consider the effect of intensity variations when predicting the loudness of the sound. Several models have been proposed to account for loudness of time-varying sounds, e.g., the model for time-varying loudness (*TVL*) by Glasberg and Moore (2002) and the dynamic loudness model (*DLM*) by Chalupper and Fastl (2002). The current DIN 45631/A1 (2010) also includes the possibility to predict loudness for time-varying signals and according to Sottek (2013) it may also be part of the

---

[1] A modified Version of this chapter is published as:

Hots, J., Rennies, J. and Verhey, J.L. **(2014)** "Modeling temporal integration of loudness" Acta Acust. united Ac. **100** (1) 184–187.

of the revised international standard for an algorithm to predict loudness (ISO 532, 1975).

A crucial prerequisite to predict loudness of time-varying intensity is a realistic simulation of the effect of duration on loudness. Loudness increases with duration, i.e., the level at equal loudness decreases as the duration increases (see, e.g., Epstein and Marozeau, 2010). This is commonly referred to as temporal integration of loudness. The results of temporal integration experiments were often described in terms of a leaky integrator with a time constant of about 100 to 200 ms (Verhey, 2010). Some studies argued that more than one time constant was involved in temporal integration. For example, Poulsen (1981) showed that the level at equal loudness as a function of duration (temporal integration function) for tone bursts can be better described by using a combination of two consecutive leaky integrators, one with a short time constant around 10 ms and one with a longer time constant around 100 ms. A combination of two leaky integrators with different time constants was also used in the model of Kumagai *et al.* (1984). In contrast to the model of Poulsen, the leaky integrators were arranged in parallel. The models by Poulsen (1981) and Kumagai *et al.* (1984) processed the intensity of the sound, i.e., they did not consider the compressive behaviour of the peripheral auditory system. A compression preceding the temporal integration stage affects the effective time constant and the shape of the predicted temporal integration function (Verhey, 2010). Compression increases the effective time constant and the slope of the temporal integration function. The influence of compression could be the reason for higher temporal integration of loudness at medium levels than at low and high levels (Epstein and Marozeau, 2010; Buus *et al.*, 1997, for a review). The current loudness models *DLM* and *TVL* include both compression and a temporal integration stage. However, Rennies *et al.* (2010) showed that the shape of the temporal integration function as predicted by *DLM* and *TVL* differed from the loudness data of Poulsen (1981) at medium levels.

The present study investigated to what extent a leaky integrator model could predict temporal integration of loudness for a 1-kHz pure tone at medium levels, where a strong influence of compression is expected. In contrast to Rennies *et al.* (2010), the data of Pedersen *et al.* (1977) (as shown by Poulsen, 1981) for a reference level of 55 dB SPL and signal durations between 5 and 320 ms were used, since the data were obtained from a very large group of subjects. For this reference level, data from 144 subjects were collected in 11 laboratories, each using one of three different measurement procedures (method of adjustment, method of constant stimulus or method of maximum likelihood). Signals were presented either via headphones or using a loudspeaker. Data from the different laboratories were similar (standard deviation of about 2 dB) and interindividual standard deviations within each laboratory were usually less than 2.5 dB. The data were compared to different models ranging from conceptual models with a temporal integration stage working on the intensity or the compressed version of the intensity (referred to as leaky integrator models) up to an extended version of the dynamic loudness model (*eDLM*, Rennies *et al.*, 2009) including stages to mimic spectral selectivity and temporal aspects such as persistence or a larger spectral loudness summation for short than long signals.

The overall goal of this study was to investigate which structure of a temporal integration stage is required to predict temporal integration data when compression is taken into account. The resulting model structure could, among other aspects, have practical implications for the calculation of loudness of time-varying technical sounds. Recently, Sottek (2013) showed a model that predicted the loudness according to DIN 45631/A1 (2010). This model contained a temporal integration stage similar to the one proposed by Kumagai *et al.* (1984).

model	equation
one LI	$\left(\frac{I}{I_0}\right) \cdot \left(1 - e^{-t/\tau}\right)$
one LI c eq. τ	$\left(\frac{I}{I_0}\right)^{\alpha} \cdot \left(1 - e^{-t/\tau}\right)$
one LI c eq. τ_{eff}	$\left(\frac{I}{I_0}\right)^{\alpha} \cdot \left(1 - e^{-\alpha t/\tau}\right)$
two LI	$\left(\frac{I}{I_0}\right) \cdot \left(1 - e^{-t/\tau_1}\right) \cdot \left(1 - e^{-t/\tau_2}\right)$
two LI c ser	$\left(\frac{I}{I_0}\right)^{\alpha} \cdot \left(1 - e^{-\alpha t/\tau_1}\right) \cdot \left(1 - e^{-\alpha t/\tau_2}\right)$
two LI c par	$\left(\frac{I}{I_0}\right)^{\alpha} \cdot \left[\left(1 - e^{-\alpha t/\tau_1}\right) + \left(1 - e^{-\alpha t/\tau_2}\right)\right] \cdot \frac{1}{2}$

Table 2.1: Leaky integrator models used in this study and the underlying equations with intensity (I), reference intensity (I_0), compression (α), time constants (τ) and stimulus duration (t)

2.2 Models

Leaky integrator

Several different simple models with a leaky integrator were tested. The first stage of each of these models was the calculation of the instantaneous intensity. For two of the leaky integrator models this was directly followed by the integration stage. The stage either consisted of a single leaky integrator ($\tau = 80\,\text{ms}$; model *one LI*) or two consecutive leaky integrators ($\tau_1 = 10\,\text{ms}$ and $\tau_2 = 50\,\text{ms}$; model *two LI*). These two were proposed by Poulsen (1981). The time constants proposed by Poulsen (1981) for the data considered here (i.e, the Pedersen *et al.*, 1977, data for 55 dB) were used for the predictions. In the other model versions, a compressive stage preceded the integration stage, i.e., the instantaneous intensity values were raised to the power of $\alpha = 0.3$ prior to temporal integration. This value of α was chosen to account for the observation that loudness approximately doubles when intensity is increased by 10 dB (Epstein and Marozeau, 2010). In total, four leaky integrator model versions including compression were used. In two of them, the integration stage was realized as a single integrator. For one model version, the time constant was the same as used in model *one LI*, i.e., $\tau = 80$ ms (*one LI c eq. τ*). By adding a compressive stage to the leaky integrator stage, the value of the time constant τ changes to an effective value $\tau_{eff} = \frac{\tau}{\alpha}$ (see Verhey, 2010), i.e., the effective time constant of this model version is longer than that of the model without compression. In addition to this model, a model (*one LI c eq. τ_{eff}*) with compression and one leaky integrator was used where the time constant of the leaky integrator was set to $\tau = 24$ ms, resulting in the same *effective* time constant as in the linear model *one LI*. In the other two leaky integrator model versions, the integration stage was realized as a combination of two integrators, one with a short and one with a long time constant. In one of them (*two LI c ser*), the integrators were serially arranged. In the other model version (*two LI c par*), they were arranged in parallel, as proposed by Kumagai *et al.* (1984). The output of the model *two LI c par* was set to the mean of the two integrators. In these two model versions, the time constants were fitted to the data. In all leaky integrator model versions, the overall loudness was determined by the maximum of the output of the integration stage, as often assumed for the prediction of loudness of short sound bursts in time-varying loudness models (Rennies *et al.*, 2010). The underlying equations of the six leaky integrator models are summarized in Table 2.1.

Dynamic loudness model

The basic structure of the *eDLM* proposed by Rennies *et al.* (2009) is the same as the one of the Dynamic Loudness Model (*DLM*, Chalupper and Fastl, 2002), i.e.: To account for the lower limit of the audible frequency range, a Butterworth filter with a cut-off frequency of 50 Hz is used to high-pass filter the input time signal. The frequency selectivity of the auditory system is accounted for by analyzing the signal with a bank of 24 overlapping filters with center frequencies from 50 Hz (0.5 Bark) to 13 500 Hz (23.5 Bark). For each auditory filter, a temporal window with an equivalent rectangular duration (ERD) of 4 ms is temporally shifted along the signal in steps of 2 ms to compute the short-term root-mean-square (rms) value. The specific loudness is derived by applying a correction for the transmission through outer and middle ear followed by a compression ($\alpha = 0.23$). A specific loudness-time pattern $N_0(z,t)$ is generated accounting for forward masking and spectral masking. Specific loudness is then integrated along the frequency dimension resulting in the instantaneous loudness as a function of time. To account for the larger spectral loudness summation for short signals than for long signals, the *eDLM* uses a bandwidth-dependent amplification at stimulus onsets, i.e., the amplification for a short duration after stimulus onset increases with bandwidth, i.e., the larger the bandwidth of the stimulus, the higher the amplification. In the final stage of the model, the instantaneous loudness is integrated using a first-order low-pass filter with a cut-off frequency of 8 Hz (which corresponds to a leaky integrator with an effective time constant of 86.5 ms), resulting in the short-term loudness. The maximum of the short-term loudness is taken as an estimate of the overall loudness.

In this study, the influence of the temporal integration stage on the prediction of the temporal integration data was also investigated within the framework of the *eDLM*. To this end, the original version of the model was compared to a modified version (*eDLM mod*) with two parallel leaky integrators, one with a short and one with a long time constant, as proposed by Kumagai *et al.* (1984). The short-term loudness, i.e., the output of this temporal integration stage, was set to the mean of the two integrators. The time constants were fitted to the data.

2.3 Results and Discussion

Figure 2.1 shows model predictions (black symbols) together with experimental data of Pedersen *et al.* (1977) for a reference level of 55 dB SPL (gray symbols). The figure shows the levels of 1-kHz tone bursts with durations of 5 to 320 ms which are equally loud as a 1-kHz tone burst with a duration of 320 ms and a level of 55 dB SPL.

In panel A the model predictions of the single leaky integrator proposed by Poulsen (1981) are plotted as filled triangles (*one LI*). The good correspondence between data and predictions of the *one LI* model indicates that a single time constant already accounted for most aspects of the data. The root of the mean quadratic deviation between these predictions and the data was $\text{rms}_{diff} = 0.8$ dB, the maximum difference was $\text{max}_{diff} = 1.2$ dB. For short signal durations, the slope of the predicted temporal integration function was close to -3 dB per doubling and decreased towards longer durations. However, this model does not include basic characteristics of auditory models, such as compression. Open triangles show predictions of the *one LI* model including compression. The compression-related modification of the effective time constant can be observed by comparing the predictions with compression (*one LI c eq. τ*, up-pointing triangles) and without compression (filled triangles).

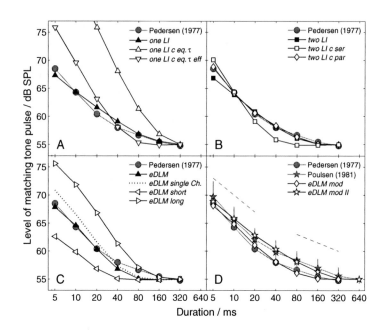

Figure 2.1: Temporal integration data and predictions. In all panels the experimental data of Pedersen *et al.* (1977) for a reference level of 55 dB are shown as gray dots. **A** Predictions of leaky integrator models with one time constant: linear (*one LI*, filled triangles), with compression (*one LI c eq. τ*, up-pointing triangles), with the same time constant as for *one LI*, and with compression and the same effective time constant (*one LI c eq. τ_{eff}*, down-pointing triangles). **B** Predictions of leaky integrator models with two time constants: linear consecutive (*two LI*, filled squares), with compression consecutive (*two LI c ser*, squares), and with compression arranged in parallel (*two LI c par*, diamonds), the latter two with optimized time constants. **C** Predictions of the *eDLM*: default parameter set (filled triangles), with a short time constant (*eDLM short*, left-pointing triangles), with a long time constant (*eDLM long*, right-pointing triangles) and using only the channel containing the signal frequency (*eDLM single ch.*, dotted line). **D** Experimental data by Poulsen (1981) (gray stars, error bars indicate 95% confidence limits) and predictions of the *eDLM* with the modified integration stage: time constants fitted to the data of Pedersen *et al.* (1977) (*eDLM mod*, diamonds) and time constants fitted to the data of Poulsen (1981) (*eDLM mod II*, stars). Dashed lines visualize slopes of -3 dB and -1.5 dB per doubling.

In addition to a change in the effective time constant, compression also increased the absolute value of the slope of the temporal integration function for short durations by $10 \log\left(\frac{1}{\alpha}\right)$ (see Verhey, 2010). Thus, even a change of the time constant of the leaky integrator to an effective time constant of $\tau_{eff} =$

80 ms (*one LI c eq.* τ_{eff}, down-pointing triangles) resulted in poor predictions (rms$_{diff}$ = 13.1 dB and max$_{diff}$ = 21.6 dB for τ = 80 ms; rms$_{diff}$ = 3.6 dB and max$_{diff}$ = 7.3 dB for τ_{eff} = 80 ms).

Panel B shows predictions of the *two LI* model proposed by Poulsen (1981) (filled squares). Using this model containing a second time constant did not lead to a substantial change in the predictions compared to the predictions of the *one LI* model (rms$_{diff}$ = 0.7 dB and max$_{diff}$ = 1.7 dB for *two LI*). This is presumably due to the choice of the short time constant which was close to the shortest duration (5 ms) used in the study. Open squares show the predictions of the *two LI* model including compression (*two LI c ser*). The time constants of this model were optimized by minimizing rms$_{diff}$, resulting in effective values of τ_1 = 2 ms and τ_2 = 47 ms. The predicted temporal integration function was steeper than the experimental data (rms$_{diff}$ = 1.3 dB, max$_{diff}$ = 2.2 dB). Especially for medium durations the level of the matching tone pulse was underestimated by up to 2.2 dB. Since the short time constant τ_1 was less than half of the minimum duration used in the experiment by Pedersen *et al.* (1977), the model is essentially a single leaky integrator model, i.e., does not benefit from the second integrator. Open diamonds show predictions of the *two LI c par* model with two integrators arranged in parallel including compression. The effective time constants were set to τ_1 = 17 ms and τ_2 = 153 ms, resulting in a good match between predictions and data for all durations (rms$_{diff}$ = 0.3 dB, max$_{diff}$ = 0.4 dB).

In panel C of Figure 2.1, predictions of the unmodified *eDLM* are shown as filled triangles. The predicted level differences between equally loud signals with the three shortest durations (5 ms, 10 ms, and 20 ms) and the longest duration (320 ms) were in agreement with the data. However, the predicted temporal integration function was steeper than in the experimental data, resulting in deviations between simulated and measured levels at intermediate durations ($\mathrm{rms}_{diff} = 0.8\,\mathrm{dB}$, $\mathrm{max}_{diff} = 1.6\,\mathrm{dB}$). This behavior is not specific for the *eDLM*. The predictions of the *eDLM* shown here were very similar to the predictions of the original *DLM* (see Rennies *et al.*, 2010) indicating that the additional stage for the duration effect in spectral loudness summation did not alter the predictions for temporal integration of loudness for tones. In addition, Rennies *et al.* (2010) showed that the predictions of the *TVL* and *DLM* were almost identical for short tone bursts, i.e., the deviations at intermediate durations are common to this type of loudness models. The reason for the slope larger than -3 dB per doubling is the compression included in the model. However, the slope was not as steep as in the simulations with one leaky integrator including compression (panel A). There are two likely processes that may account for the differences between the predictions of the *eDLM* and the simple leaky integrator models: the running temporal integration with the temporal window to estimate the time-varying intensity in each critical band and the spectral integration across critical bands. In order to assess if the latter process accounts for the deviations, levels at equal loudness were predicted without a spectral integration stage by using only the output of the critical band at the signal frequency (*eDLM single ch.*, dotted line). The slope of the predicted temporal integration function is higher than that of the complete model (*eDLM*). This indicates that, within the complete model, the excitation of off-frequency critical bands increases the loudness of signals with short durations, resulting in a reduction of the level difference between short and long signals at equal loudness. The simulations with the *eDLM* shed some doubts on the assumption that the slope of the temporal integration function directly corresponds to the compression of the auditory system (as, e.g., suggested by Buus *et al.*, 1997) since the slope is also affected by other properties of the auditory system such as its spectral resolution. The deviations between predictions and data at intermediate durations can not be overcome in this type of model by a simple change of the time constant. This is shown for a considerably shorter effective time constant of $\tau_1 = 45\,\text{ms}$ (*eDLM short*, open left-pointing triangles) and a considerably longer effective time constant of $\tau_2 = 185\,\text{ms}$ (*eDLM long*, open right-pointing triangles).

Panel D shows the predictions of the *eDLM* with a modified temporal integration stage as described in Section 2.2. The time constants of the two parallel leaky integrators of the modified temporal integration stage were optimized in the same way as for the other model versions with optimized time constants. The predictions of this optimized model with effective time constants of $\tau_1 = 45\,\text{ms}$ and $\tau_2 = 185\,\text{ms}$ (*eDLM mod*) are shown as diamonds. The modification of the *eDLM* led to a good prediction of the data for the whole range of durations ($\text{rms}_{diff} = 0.5\,\text{dB}$, $\text{max}_{diff} = 0.7\,\text{dB}$). For medium to long durations this model structure predicted a slope of -1.5 dB per doubling, as it would also emerge from a multiple looks strategy (Viemeister and Wakefield, 1991). For short durations the slope approached -3 dB per doubling (slopes are visualized by dashed lines in the bottom right panel). A model combining a leaky integrator with a short time constant for short durations and a multiple looks strategy for longer duration was used in Oxenham *et al.* (1997) to predict the influence of duration on detection of a masked tone. The present approach shows that a similar result can be achieved by using two parallel leaky integrators.

The good prediction is primarily due to the parallel arrangement of the leaky integrators. If only one of the two leaky integrators was used, the slope would be too steep, as shown in panel C of Fig. 2.1. The slope was even higher for two consecutive integrators. Gray stars in panel D also show data of Poulsen (1981) at a reference level of 55 dB SPL. Error bars indicate 95% confidence limits. Compared to Pedersen *et al.* (1977), Poulsen (1981) measured a temporal integration function for a slightly larger range of durations, but a smaller group of subjects. Deviations between the data of Pedersen *et al.* (1977) and Poulsen (1981) are apparent. It should be noted that the deviations between the two sets of data are smaller than the deviation between the data from the 11 individual laboratories summarized in Pedersen *et al.* (1977). Using the same parameters for the temporal stage of the *eDLMmod* as for the Pedersen data, the rms_{diff} was 1.4 dB with a maximum difference max_{diff} of 2.3 dB. This shows the accuracy of this model to predict other data of temporal integration with a smaller set of subjects. Using the same approach as in Poulsen (1981), i.e., allowing individual time constants for each data set, a good agreement between data and predictions with optimized time constants ($\tau_1 = 35$ ms and $\tau_1 = 398$ ms) could be obtained (stars in panel D). Thus, the present approach overcomes the problem of too steep predicted temporal integration curves in models with cochlear compression by using a parallel organization of leaky integrators with different time constants. The assumption that more than one process is involved in temporal integration was made in several previous studies, but these usually assumed, in contrast to the present study, a sequential organization of these stages. The present study showed that such a serial organization of leaky integrators results in predicted temporal integration functions, which are steeper than observed in the data, when compression is taken into account. The improvement of temporal integration by using the modified temporal integration stage was demonstrated within the model framework of the *eDLM* (Rennies *et al.*, 2009), but it is likely that a similar change of the stage calculating the short-term loudness in the *TVL* of Glasberg and Moore (2002) would lead to a similar improvement in prediction accuracy for the temporal integration data considered here.

In summary, the results of the present study indicate that a certain structure of the temporal integration stage (two leaky integrators in parallel) is crucial to predict temporal integration over the whole range of durations. The time constants were fitted to the presumably largest set of temporal integration data for loudness, but for other studies slightly different time constants may better fit the data. Interestingly, the model recently presented by Sottek (2013) which fulfilled the requirements of the DIN 45631/A1 (2010) concerning the instantaneous loudness functions shown in Appendix C included a temporal integration stage very similar to the one proposed here. This study shows that such a structure of the temporal integration stage is indispensable when predicting temporal integration over a large range of durations.

3 Loudness of sounds with a subcritical bandwidth: a challenge to current loudness models?[1]

3.1 Introduction

Loudness as the primary perceptual correlate of the level of a stimulus also changes with other physical stimulus properties (see, e.g., Florentine, 2011). One of these properties is the stimulus bandwidth. In general, loudness increases with increasing bandwidth, an effect commonly referred to as spectral loudness summation (e.g., Zwicker *et al.*, 1957; Scharf, 1978). This effect, however, is only observed when the bandwidth exceeds a critical bandwidth. For bandwidths smaller than the critical bandwidth data suggest that loudness does not vary with bandwidth (e.g., Zwicker *et al.*, 1957). However, for very narrow bands of noise the inherent envelope fluctuations may affect loudness, as

---

[1] A modified Version of this chapter is published as:

Hots, J., Rennies, J. and Verhey, J.L. **(2013)** "Loudness of sounds with a subcritical bandwidth: a challenge to current loudness models?" J. Acoust. Soc. Am. **134** (4) EL334–EL339.

pointed out by Scharf (1978). The effect of level fluctuations on loudness was usually investigated using amplitude modulated sounds. It was found that modulation can increase loudness at a constant RMS level for certain combinations of modulation frequency and carrier bandwidth (e.g., Zhang and Zeng, 1997; Moore *et al.*, 1999). Thus, one may expect a higher loudness for a noise with a subcritical bandwidth than a pure tone at the center frequency of the noise as long as the auditory system is sensitive to the inherent level fluctuations. To test this hypothesis, this study measured level differences at equal loudness between bandpass noises centered at 1.5 kHz and a 1.5-kHz pure tone using an adaptive loudness matching procedure. To rule out bias effects due to the choice of the reference, the level differences at equal loudness were measured using two references. Data were compared with predictions of current dynamic loudness models.

3.2 Methods

A 1.5-kHz pure tone and bandpass filtered Gaussian noise stimuli geometrically centered at 1.5 kHz with bandwidths of 5, 15, 45, 135, and 405 Hz were matched in loudness using an adaptive two-interval, two-alternative forced-choice procedure with interleaved tracks. In each trial, a test and a reference stimulus were presented in random order, separated by a silent interval of 500 ms. The participants indicated which interval they perceived as the louder one by pressing the corresponding key. The level of the test stimulus was varied according to a one-up one-down rule, starting with a step size of 8 dB, which was halved after every upper reversal until a step size of 2 dB was reached. With this step size the track continued for four reversals. To estimate the level at equal loudness, the mean of these four last reversals was calculated. For each condition, starting levels of -10 dB, 0 dB and +10 dB relative to the reference level of 50 dB SPL were used. The experiment was divided into two runs. In the first run (tone reference), the reference was the 1.5-kHz pure tone. In the second run (noise reference), 135-Hz wide noise stimuli were used as the reference. No training trials were given prior to the measurements. All stimuli were generated digitally at a sampling frequency of 44.1 kHz, had a duration of 500 ms, and were gated using a 5-ms $\cos^2$ window. A new token of random noise was generated for each presentation. Stimulus generation and presentation as well as the recording of the results were controlled using MATLAB. Stimuli were D/A converted (RME ADI-8 PRO) and presented diotically via an attenuator (Tucker-Davis HB7) and headphones (Sennheiser HD650) in a double walled sound-proof booth. Ten normal-hearing listeners (six male, four female) aged from 23 to 27 years participated in the experiment. They had thresholds in quiet $\leq$ 15 dB HL at standard audiometric frequencies between 125 Hz and 8 kHz. All participants had previous experiences with psychoacoustic measurements and were paid unless they were members of the workgroup.

The experiments were simulated with the model for time-varying loudness (*TVL*) of Glasberg and Moore (2002) and the extended dynamic loudness model (*eDLM*) proposed by Rennies *et al.* (2009). The *TVL* consists of the following stages: (i) bandpass filter simulating the outer and middle ear transfer function, (ii) six parallel Fast Fourier Transforms to generate excitation patterns, (iii) transformation to instantaneous loudness, (iv) transformation to short-term loudness with an integrator (time constants: 22 ms for attack, 50 ms for release), and (v) integration with longer time constants (99 ms for attack, 2000 ms for release) to derive long-term loudness. The maximum of the short-term loudness and the mean of the long-term loudness were taken as estimates of the overall loudness. The *eDLM* consists of the following stages: (i) 50-Hz high-pass filter to account for the lower limit of the audible frequency range, (ii) filtering with 24 overlapping filters with center frequencies from 50 Hz to 13 500 Hz, (iii) calculation of the short-term root-mean-square with a sliding temporal window (equivalent rectangular duration of 4 ms), (iv) correction for the transmission through outer and middle ear followed by compression, (v) transformation to specific loudness-time pattern accounting for forward masking and spectral masking, (vi) integration across frequency, (vii) bandwidth-dependent amplification at stimulus onsets, and (viii) low-pass filtering (cut-off frequency of 8 Hz) to account for temporal integration of loudness. The maximum of the resulting short-term loudness was taken as an estimate of the overall loudness.

3.3 Results and Discussion

Figure 3.1 shows the mean experimental data for level differences at equal loudness between noise and pure tone as a function of noise width of the 10 participants (individual results are shown in Fig. 8.1 in the Appendix 8.1). Error bars indicate the interindividual standard deviation. For the tone reference (open circles), the measured level differences at equal loudness were close to zero for bandwidths of 5 and 15 Hz. For larger bandwidths, the noise required a higher level to be perceived as equally loud compared to a pure tone. This level difference increased with bandwidth up to about 5 dB at 135 Hz. For the noise reference (open squares), the level differences at equal loudness were transformed to the scale used for the other data set ($L_{noise} - L_{tone}$) by subtracting the measured (negative) level difference for the pure tone from all data points for the noises. The absolute level difference between the equally loud pure tone and the 135-Hz wide noise was about 6 dB (see data point at 135 Hz). The standard deviation for this data point is the standard deviation for the comparison of the noise with the pure tone. For all other data points, the standard de-

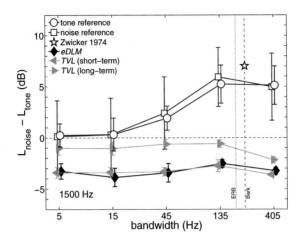

Figure 3.1: Level difference at equal loudness between a noise centered at 1.5 kHz and a 1.5-kHz pure tone as a function of the bandwidth of the noise. Open symbols: mean experimental results of the 10 participants for the tone reference (circles) and the noise reference (squares) condition. Error bars indicate interindividual standard deviations. The star shows a result of Zwicker (1974). Filled symbols: mean model predictions of 10 runs of the simulation for the *eDLM* (diamonds) and for the *TVL* on the basis of the short-term loudness (left-pointing triangles) and the long-term loudness (right-pointing triangles). Error bars indicate the standard deviation of the 10 runs. Vertical lines show the critical bandwidth at a frequency of 1.5 kHz in ERB_N (dotted line) and Bark (dash-dotted line). To increase readability data are slightly shifted against each other.

viation is that of the transformed level differences, i.e., it was calculated by taking the square root of the sum of the variance for this bandwidth and the variance for the pure tone signal. As a consequence, the standard deviations were larger for the noise reference than for the pure tone reference. Otherwise the results for the two references were the same, i.e., they differed by less than 1 dB. For the 405-Hz wide noise, no further increase was observed in the two data sets. This is presumably due to spectral loudness summation since 405 Hz is already broader than the bandwidth of the auditory filter at this frequency. According to the ERB_N scale the critical bandwidth is 186.7 Hz (dotted line in Fig. 3.1). For the Bark scale it is 224.4 Hz (dash-dotted line in Fig. 3.1). Note that even for this supercritical bandwidth of 405 Hz the level at equal loudness is still about 5 dB higher than that of a pure tone.

Data were statistically analyzed in SPSS using a one-way analysis of variances (ANOVA) for each of the two references. For both, bandwidth had a significant effect on the level difference at equal loudness (pure tone reference: $F(4) = 45.990$, $p \leq 0.001$; noise reference: $F(4) = 12.307$, $p \leq 0.001$). Post-hoc t-tests with a significance level of $p = 0.05$ using a Bonferroni correction for multiple comparisons showed the following significant differences: For the pure tone reference, the level differences for 135 and 405 Hz were significantly different from those for the other three bandwidths ($p \leq 0.001$ for all comparisons except for the one between 405 and 45 Hz, where it was $p \leq 0.01$). In addition, the level difference for the 45-Hz wide noise was significantly different from that for the 15-Hz wide noise ($p \leq 0.05$). For the noise reference, the level difference for 405 Hz was significantly higher than that for the pure tone ($p \leq 0.05$), the 5-Hz ($p \leq 0.01$), and the 15-Hz ($p \leq 0.01$) wide noises. In addition, the level for the 45-Hz wide noise was significantly different from that for the 5-Hz wide noise ($p \leq 0.05$).

The data presented here are neither in line with data from other studies (e.g., Zwicker *et al.*, 1957; Zhang and Zeng, 1997; Moore *et al.*, 1999), which found little or no effect of slow amplitude modulations on loudness, nor with stationary loudness models predicting no level difference of equally loud subcritical noise and pure tones at the same level. However, other studies, some using different measurement procedures, also showed positive level differences between noises and equally loud tones, but these contradicting results were generally not discussed. For example, a positive level difference between a critical-band wide noise and an equally loud pure tone was measured by Zwicker (1974) for center frequencies of 1, 1.5, and 2 kHz and levels of 30, 50, 70, and 90 dB. His data point with comparable center frequency and level is shown as star in Fig. 3.1. The level difference was about 7 dB, i.e., similar to the maximum level difference found in the present study. An effect in the same direction as found in the present study was also reported by Grimm *et al.* (2002). In contrast to the present study, they used a 2-kHz center frequency, a bandwidth of 32 Hz, and a reference level of 65 dB. Furthermore they generated the bandpass noise by multiplying a pure tone with a low-pass filtered noise, i.e., they used multiplied noise, whereas the present study used Gaussian noise. Despite the difference in stimulus parameters, very similar results were obtained: Grimm *et al.* (2002) measured a level difference at equal loudness of 1.5 dB between bandpass noise and a pure tone while the present study obtained 2 dB for the 45-Hz wide noise. A 2-dB difference was also found by Florentine *et al.* (1978) for a 1-kHz pure tone and an equally loud 220-Hz wide noise. An effect of bandwidth on loudness similar to the one found in the present study can be observed in the data of Zwicker *et al.* (1957) for a 210-Hz wide noise reference centered at 1420 Hz and a signal level of 50 dB. For their test signal with the narrowest bandwidth (35 Hz), the level of the reference had to be about 2 dB higher than that

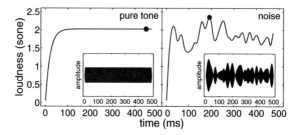

Figure 3.2: Loudness N(t) of a pure tone at a frequency of 1.5 kHz
(left panel) and an equal-level bandpass filtered noise
centered at 1.5 kHz with a bandwidth of 45 Hz (right
panel) calculated by the *eDLM*. Filled dots show the
maximum of the loudness-time function. Insets in each
panel show the time waveform of the corresponding signal.

of the test signal to be perceived as equally loud. The level
difference decreased to 0 dB at a bandwidth of 150 Hz.

Figure 3.1 also shows model predictions. The levels at equal
loudness were determined for 10 realizations of the noise. The
mean and standard deviations of these 10 estimates of the level
difference at equal loudness are shown for the *eDLM* (black diamonds) and the *TVL* based on short-term loudness (gray left-
pointing triangles). Both models predict level differences between about -2.5 dB and -4 dB for the subcritical bandwidths,
i.e., a higher level for the tone at equal loudness, in contrast
to the data. For the bandwidth of 405 Hz the predictions are
similar to those for 135 Hz. Figure 3.2 shows loudness time functions calculated with the *eDLM* for a 1.5-kHz pure tone at a
level of 50 dB SPL (left panel) and for a bandpass filtered noise
signal centered at 1.5 kHz with a bandwidth of 45 Hz at the

same level (right panel) to illustrate why the models predict such a negative level difference. The insets in each panel show the time waveform of the corresponding signal. The loudness time function follows the inherent level fluctuations of the narrowband noise. Using the maximum (dots in Fig. 3.2) as an estimate of the overall loudness thus yields a higher predicted loudness for subcritical noises than for pure tones at the same level. The long-term loudness in the *TVL* is specifically intended to predict the loudness of fluctuating sounds. Based on this loudness, the model predicts level differences between about -0.5 and -1 dB for the subcritical bandwidths (gray right-pointing triangles in Fig. 3.1), which is not in agreement with the present data for larger bandwidths, but is close to results of previous studies not showing an effect of bandwidths in the subcritical range (see, e.g., Zwicker *et al.*, 1957). Note that these simulations argue against the hypothesis that the failure to predict the effect is mainly due to the estimation of the overall loudness on the basis of the maximum, since here the temporal average is taken as an estimate of the overall loudness.

One may argue that the positive level difference may result from the effect of peripheral compression on signals with level fluctuations. For example, Buschermöhle *et al.* (2007) showed that the average compressed envelope of a signal consisting of several comodulated narrowband noises (resulting in large level fluctuations) is smaller than that of a stimulus with the same spectrum but with uncorrelated level fluctuations of the noise bands. An argument against compression as an explanation for the results is that it would not predict the bandwidth-dependent level difference observed in the data.

The positive level difference observed in the data may be related to another aspect of the nonlinear properties of the auditory periphery, namely suppression. Several studies comparing spectral selectivity in simultaneous masking and forward masking experiments indicate that the effective auditory filter width is narrower for the latter (e.g., Moore and Glasberg, 1981; Buchholz *et al.*, 2010). This effect was attributed to suppression, which is absent in forward masking conditions but reduces the signal representation in simultaneous masking conditions. In this light, the decrease in loudness as the bandwidth is increased from 15 to 135 Hz may reflect an increase in the magnitude of suppression, since an increasing portion of the stimulus falls into suppressive regions of the auditory filter.

For low modulation frequencies, Zhang and Zeng (1997) found an effect of 2 dB in the opposite direction for three-tone complexes only, while for two-tone complexes the effect was nearly absent. In the present study there was no level difference between an equally loud pure tone and noises up to a bandwidth of 15 Hz. As for the noise the frequency components of the two-tone complex have equal intensity. Thus, the data for the two-tone complexes may be more comparable to the data presented in this study. In contrast, the levels of the sidebands are 6 dB lower than that of the carrier for the three-tone complex.

According to the suppression hypothesis, one has to assume on the basis of the current data that suppression starts to affect signal representation for bandwidths as narrow as 45 Hz. This is considerably smaller than the factor of 1.2 to 1.8 usually found for the ratio between filters derived from simultaneous and forward masking data (Moore and Glasberg, 1981; Buchholz *et al.*, 2010). Further studies are required to test the hypothesis that the suppression contributed to the observed positive level differences.

4 Loudness of subcritical sounds as a function of bandwidth, center frequency, and level[1]

4.1 Introduction

Loudness increases with increasing bandwidth as soon as a certain bandwidth, referred to as critical bandwidth, is exceeded (e.g., Zwicker *et al.*, 1957; Scharf, 1978). This effect is commonly known as spectral loudness summation. It can be modeled by assuming a critical band filtering followed by a compression and summation across critical bands (DIN 45631/A1, 2010; ANSI S3.4, 2007; Fastl and Zwicker, 2007). For sounds with bandwidths narrower than this critical bandwidth, such an approach predicts a loudness that does not depend on bandwidth. For very narrow bandwidths, however, inherent level fluctuations may affect the loudness of noise signals (Scharf, 1978).

---

[1] A modified Version of this chapter is published as:
Hots, J., Rennies, J. and Verhey, J.L. "Loudness of subcritical sounds as a function of bandwidth, center frequency, and level" J. Acoust. Soc. Am. **135** (3) 1313–1320.

Several studies on loudness of amplitude-modulated sounds showed that amplitude modulation affects loudness, at least at low modulation frequencies. Most of them seem to suggest a slight increase in loudness after imposing an amplitude modulation although this seems to depend on stimulus parameters (see, e.g., Bauch, 1956; Moore *et al.*, 1999; Zhang and Zeng, 1997). To account for the higher loudness of amplitude-modulated tones, dynamic loudness models either use an appropriate decision variable, like the maximum of the short-term loudness as in the dynamic loudness model (DLM, Chalupper and Fastl, 2002; Rennies *et al.*, 2009) or by an temporal integration stage with different attack and release time as in the loudness model for time-varying sounds (TVL, Glasberg and Moore, 2002).

In Chapter 3 these models were used to predict the level difference at equal loudness of tones and noises with subcritical bandwidth. It was shown that TVL and DLM predicted a negative level difference, i.e., a higher level for the tone. In contrast to these predictions, participants measured a positive level difference at equal loudness which increased with bandwidth up to the critical bandwidth, suggesting that the model assumptions by stationary as well as dynamic loudness models are not sufficient to account for loudness of subcritical noise bands. So far this effect has only been shown using a matching procedure for a center frequency of 1.5 kHz at a reference level of 50 dB SPL. The present study extends the data investigating how the reference level and the center frequency influence the effect. In addition it is tested if the effect is also observed when categorical loudness scaling is used instead of a matching procedure.

4.2 Experiment 1: Effect of bandwidth using a matching procedure

4.2.1 Methods

Gaussian noise stimuli geometrically centered at 750 or 3000 Hz with bandwidths of 5, 15, 45, 135, 405, 810 and 1620 Hz were used as test stimuli. A pure tone at the corresponding center frequency at a reference level of 50 dB SPL served as a reference stimulus. An adaptive two-interval, two-alternative forced-choice procedure with interleaved tracks was used to measure the level difference at equal loudness between reference and test stimuli. In each trial, test and reference stimulus were presented in random order, separated by a silent interval of 500 ms. By pressing the corresponding key the participants indicated which interval they perceived as the louder one. The level of the test stimulus was varied according to a one-up one-down rule. The step size was 8 dB at the beginning of the track. It was halved after every upper reversal until a step size of 2 dB was reached. With this step size the track continued for another four reversals. The mean of these four last reversals was calculated to estimate the level at equal loudness. For each test stimulus, there were three tracks, differing in their starting level (-10 dB, 0 dB or +10 dB relative to the reference level). The mean of the three estimates was taken as the final estimate of the level difference at equal loudness for the test stimulus. For each center frequency, the tracks for the different bandwidths and starting levels were interleaved.

All stimuli were generated digitally at a sampling frequency of 44.1 kHz, had a duration of 500 ms, and were gated using a 5 ms $\cos^2$ window. A new token of random noise was generated for each presentation. Stimulus generation and presentation as well as the recording of the results were controlled using the MATLAB AFC framework by Ewert (2013). Stimuli were D/A converted (RME ADI-8 PRO) and presented diotically via an attenuator (Tucker-Davis HB7) and headphones (Sennheiser HD650) in a double walled sound-proof booth. Nine normal-hearing listeners (five male, four female) aged from 23 to 27 years participated in the experiment. Their thresholds in quiet were ≤ 15 dB HL at the standard audiometrical frequencies between 125 Hz and 8 kHz. All participants had previous experiences with psychoacoustic measurements and had also participated in the study shown in Chapter 3. They were paid unless they were members of the workgroup.

4.2.2 Results

Figure 4.1 shows individual level differences at equal loudness between noise centered at 750 Hz and the 750-Hz pure tone as a function of noise bandwidth for the nine participants (individual results for the center frequency of 3000 Hz are shown in Fig. 8.2 in the Appendix 8.2). Each panel shows individual data of one of the nine subjects. Error bars indicate individual standard deviations of the three repetitions of the experiment. For eight of the nine participants the standard deviation is rather small (usually less than 2 dB) whereas for one participant (P7) it has larger values up to about 6.5 dB at a bandwidth of 810 Hz. Vertical lines indicate the critical bandwidth at a frequency of 750 Hz in equivalent rectangular bandwidth (ERB_N, Moore and Glasberg, 1981, dotted line) and Bark (Fastl and Zwicker, 2007, dash-dotted line). All participants measured a positive level difference at equal loudness at intermediate bandwidths with a maximum at 135 or 405 Hz. The size of the effect differed across subjects. At 135 Hz, the level difference ranged from 3 dB (P2 and P3) to 12.5 dB (P6). The mean data across all participants for 750 Hz is shown in panel A of Fig. 4.2. Panel B shows the mean data for the same set of participants for the center frequency of 3000 Hz. As in Fig. 4.1 the level difference at equal

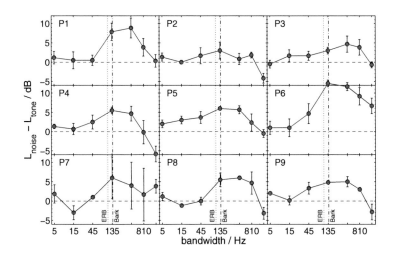

Figure 4.1: Individual results of the nine participants in Experiment 1 for signals centered at 750 Hz. Level difference at equal loudness between noise and a pure tone as a function of the bandwidth of the noise at a reference level of 50 dB SPL. Error bars indicate intraindividual standard deviations. In each panel vertical lines show the critical bandwidth at a frequency of 750 Hz in ERB_N (dotted line) and Bark (dash-dotted line).

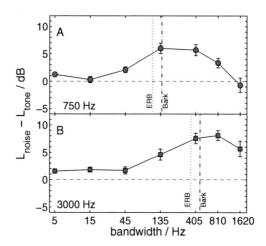

Figure 4.2: Mean results of Experiment 1. Level difference at equal loudness between noise and a pure tone as a function of the bandwidth of the noise at a reference level of 50 dB SPL for signals centered at 750 Hz (panel A, circles) and 3000 Hz (panel B, squares). Error bars indicate interindividual standard errors. In each panel vertical lines show the critical bandwidth at the corresponding center frequency in ERB_N (dotted line) and Bark (dash-dotted line).

loudness is shown as a function of noise bandwidth. Error bars indicate the interindividual standard error. Vertical lines again indicate the critical bandwidth at the corresponding center frequency in ERB_N (dotted line) and Bark (dash-dotted line). For the center frequency of 750 Hz, the level difference at equal loudness is close to zero for the two narrowest noise signals. As the bandwidth increases, the level difference at equal loudness increases up to the bandwidth of 135 Hz where it is about 6 dB. For broader noises, it decreases with increasing bandwidth. At the bandwidth of 1620 Hz it has a value of about -1 dB. Comparable results are obtained for the center frequency of 3000 Hz (panel B in Fig. 4.2): For bandwidths from 5 Hz to 45 Hz the level difference is small (1 to 2 dB). For broader noises it increases up to the bandwidth of 810 Hz where it has a value of about 8 dB. For the bandwidth of 1620 Hz it decreases to a value of about 5.5 dB.

The data were analyzed using a two-way analysis of variances (ANOVA) on the factors center frequency and noise bandwidth in SPSS. A significant effect of center frequency ($F(1) = 8.514$, $p \leq 0.05$) and noise bandwidth ($F(6) = 19.219$, $p \leq 0.001$) as well as of their interaction ($F(6) = 13.373$, $p \leq 0.001$) was found. A following one-way analysis of variances (ANOVA) was used to analyze the influence of the bandwidth for each of the two center frequencies separately. For both center frequencies, the bandwidth had a significant effect on the level difference at equal loudness (750 Hz: $F(6) = 14.671$, $p \leq 0.001$; 3000 Hz: $F(6) = 20.133$, $p \leq 0.001$). Post-hoc t-tests with a significance level of $p = 0.05$ using a Bonferroni correction for multiple comparisons showed the following significant differences: For the center frequency of 750 Hz, the level difference for 135 Hz was significantly different from those of 5, 15, and 45 Hz ($p \leq 0.05$) and 1620 Hz ($p \leq 0.01$). The level difference for 405 Hz was significantly different from those of 15 Hz ($p \leq 0.05$) and 1620 Hz ($p \leq 0.01$). For the center frequency of 3000 Hz, the level differences for 5, 15, and 45 Hz were significantly different from those of 405 and 810 Hz ($p \leq 0.001$ for the comparisons of 5 Hz with 810 Hz and 15 Hz with 405 and 810 Hz; $p \leq 0.01$ for all other comparisons). In addition, the level difference for the 135-Hz wide noise was significantly different from that for the 810-Hz wide noise ($p \leq 0.05$).

4.3 Experiment 2: Effect of level using a matching procedure

4.3.1 Methods

The measurement procedure, stimuli generation and presentation as well as the recording of the results in this experiment were the same as in Experiment 1 (see Sec. 4.2.1). Gaussian noise stimuli with bandwidths of 15, 135, 405, and 1215 Hz were matched in loudness to a pure tone for three center frequencies: 750, 1500, and 3000 Hz. Three reference levels were used for each center frequency: 30, 50, and 70 dB SPL. For each center frequency and level, the tracks for the different bandwidths and starting levels were interleaved separately. The order of the measurements for the different center frequencies and reference levels was randomized. The setup including the laboratory space differed from those used in Experiment 1. Stimuli were D/A converted and presented diotically via an external sound card (RME Fireface 400) and headphones (Sennheiser HD650). Ten normal-hearing listeners (four male, six female) aged from 21 to 38 years participated in the experiment. They had thresholds in quiet ≤ 15 dB HL at standard audiometrical frequencies between 125 Hz and 8 kHz. Except for two participants, all had previous experiences with psychoacoustic measurements. Only two of the 10 participants took part in Experiment 1. All participants were paid unless they were members of the workgroup.

4.3.2 Results

Figure 4.3 shows mean level differences at equal loudness between noise and pure tone stimuli as a function of the noise bandwidth for center frequencies of 750 Hz (panel A, circles), 1500 Hz (panel B, triangles), and 3000 Hz (panel C, squares) at the reference levels of 30 dB (black filled symbols), 50 dB (gray filled symbols), and 70 dB (open symbols) (individual results of the 10 participants are shown in Fig. 8.3 to Fig. 8.5 in the Appendix 8.2). As in Experiment 1, for all center frequencies and reference levels a positive level difference was measured at equal loudness at intermediate bandwidths. For the narrowest bandwidth of 15 Hz the measured level difference at equal loudness was close to 0 dB (from about 0.5 dB for 750 Hz at a reference level of 30 dB to 2 dB for 3000 Hz at a reference level of 50 dB). With increasing bandwidth the level difference increased up to a peak at a bandwidth close to the critical bandwidth. The maximum positive level difference of about 8 dB was obtained for a center frequency of 3000 Hz at a reference level of 30 dB and a bandwidth of 405 Hz. For bandwidths larger than the critical bandwidth, level difference decreased with bandwidth.

The effect of a positive level difference at equal loudness decreases with increasing reference level. It is largest for the reference level of 30 dB and smallest for the reference level of 70 dB. A two-way analysis of variances (ANOVA) on the factors reference level and noise bandwidth, which was performed for each of the center frequencies separately showed a significant effect of level (750 Hz: $F(2) = 11.825$, $p \leq 0.001$; 1500 Hz: $F(2) = 3.689$, $p \leq 0.05$; 3000 Hz: $F(1.270) = 10.626$, $p \leq 0.01$; Greenhouse Geisser used at 3000 Hz) and bandwidth (750 Hz: $F(3) = 62.495$, $p \leq 0.001$; 1500 Hz: $F(1.350) = 8.630$, $p \leq 0.01$; 3000 Hz: $F(1.555) = 6.986$, $p \leq 0.05$; Greenhouse Geisser used at 1500 and 3000 Hz) for all center frequencies. A significant effect was found for the interaction of these two factors at

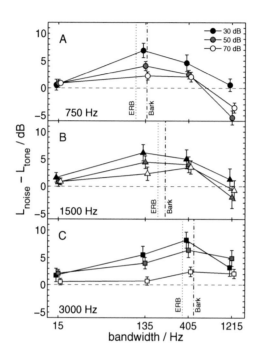

Figure 4.3: Mean results of Experiment 2. Level difference at equal
loudness between noise and a pure tone as a function
of the bandwidth of the noise for signals centered at
750 Hz (panel A, circles), 1500 Hz (panel B, triangles)
and 3000 Hz (panel C, squares) for reference levels of
30 dB SPL (black filled symbols), 50 dB SPL (gray filled
symbols) and 70 dB SPL (open symbols). Error bars
indicate interindividual standard errors. In each panel
vertical lines show the critical bandwidth at the corresponding center frequency in ERB_N (dotted line) and
Bark (dash-dotted line). To increase readability data
are slightly shifted against each other.

750 Hz and 3000 Hz (750 Hz: $F(6) = 9.881$, $p \leq 0.001$; 3000 Hz: $F(6) = 4.877$, $p \leq 0.001$). For the center frequency of 1500 Hz this effect was not significant ($F(6) = 1.766$, $p = 0.124$). For the center frequencies of 750 Hz and 3000 Hz one-way analyses of variances (ANOVA) were used to further analyze the influence of the reference level for each of the bandwidths separately. These showed a significant influence of the reference level at the bandwidth of 135 Hz ($F(2) = 13.303$, $p \leq 0.001$ for the center frequency of 750 Hz; $F(2) = 13.872$, $p \leq 0.001$ for the center frequency of 3000 Hz), for the bandwidth of 1215 Hz at the center frequency of 750 Hz ($F(2) = 17.226$, $p \leq 0.001$) and for the bandwidth of 405 Hz at the center frequency of 3000 Hz ($F(2) = 12.343$, $p \leq 0.001$). For the conditions showing a significant effect of the reference level, post-hoc t-tests with a significance level of $p = 0.05$ using a Bonferroni correction for multiple comparisons show a significant difference between the reference levels of 30 dB and 70 dB ($p \leq 0.01$ for the center frequency of 750 Hz at the bandwidth of 135 Hz and for the center frequency of 3000 Hz at the bandwidth of 135 Hz; $p \leq 0.05$ for the center frequency of 750 Hz at the bandwidth of 1215 Hz and for the center frequency of 3000 Hz at the bandwidth of 405 Hz), between the reference levels of 50 dB and 70 dB ($p \leq 0.05$ for the center frequency of 750 Hz at the bandwidth of 135 Hz and for the center frequency at the bandwidths of 135 Hz and 405 Hz), and between the reference levels of 30 dB and 50 dB at the center frequency of 750 Hz and the bandwidth of 1215 Hz ($p \leq 0.001$).

4.4 Experiment 3: Subcritical loudness using categorical loudness scaling

4.4.1 Methods

For a subset of the stimuli used in Experiment 2, loudness functions were measured using an adaptive categorical loudness scaling procedure (Brand and Hohmann, 2002). The procedure fulfills the requirements of the DIN ISO 16832 (2007). The subset contained a pure tone and a Gaussian noise geometrically centered at the tone frequency for each of the three center frequencies of 750, 1000, and 3000 Hz. The bandwidth of the noise stimuli was chosen to elicit a large effect, i.e., close to the respective critical bandwidth: It was 135 Hz for center frequencies of 750 and 1500 Hz, and 405 Hz for a center frequency of 3000 Hz. Stimuli were presented at various presentation levels covering the entire dynamic range of the individual participant, which was determined in the first phase of the experiment (Brand and Hohmann, 2002). The task of the participants was to rate the loudness on a scale of 11 loudness categories. The scale included two extreme categories "unhörbar" ("inaudible") and "extrem laut" ("extremely loud") and the five named categories "sehr leise" ("very soft"), "leise" ("soft"), "mittel" ("medium"), "laut" ("loud"), and "sehr laut"("very loud"). In addition, four unnamed intermediate categories could be selected. The measurements for the different stimuli were conducted consecutively in randomized order. The measurement was repeated three times by each participant. Stimulus generation and presentation as well as the recording of the results were controlled using MATLAB. The measurement setup was the same as used in Experiment 2 and the same 10 normal-hearing listeners participated (see Sec. 4.3.1). For an analysis of the data, numerical values (categorical units, cu) between 0 (inaudible) and 50 (extremely loud) have been linearly assigned to the loudness categories. To receive individual loudness functions for each participant and track, a model function as suggested by Brand and Hohmann (2002) was fitted to the responses of all three repetitions of the measurement. The experiment was performed in a running-noise condition, i.e., a new token of random noise was generated for each presentation and in a frozen-noise condition, i.e., a new token of random noise was generated in the beginning of each of the three repetitions of a track and used for each presentation within this repetition of the track.

4.4.2 Results

Figure 4.4 shows individual loudness functions of the running noise condition of the experiment for 750-Hz pure tones (solid lines) and the Gaussian noise stimuli geometrically centered at 750 Hz with a bandwidth of 135 Hz (dashed lines) for the 10 participants (individual loudness functions for the running noise condition and the other center frequencies are shown in Fig. 8.6 to Fig. 8.10 in the Appendix 8.2). For levels of the pure tone of 30, 50, and 70 dB, the level differences at equal loudness between the noise and the pure tone were calculated from these loudness functions. Figure 4.5 shows mean values of the level difference at equal loudness as a function of level for center frequencies of 750 Hz (panel A), 1500 Hz (panel B), and 3000 Hz (panel C) for the running noise condition (stars) and the frozen noise condition (diamonds). Error bars indicate interindividual standard errors. The level differences at equal loudness have values between about 3.5 and 6 dB at a level of 30 dB, about 3 and 5 dB at a level of 50 dB, and between about 1.5 and 4 dB at a level of 70 dB showing a trend of decreasing level differences at equal loudness with increasing level. For comparison, Fig. 4.5 also shows the results at the corresponding bandwidths of Experiment 2 (circles in panel A, triangles in panel B, and squares in panel C). For all center

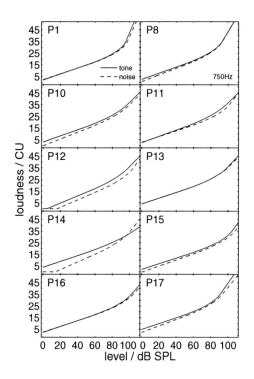

Figure 4.4: Mean loudness functions of the 10 participants in Experiment 3 for pure tones (solid lines) and running noise signals (dashed lines) centered at 750 Hz.

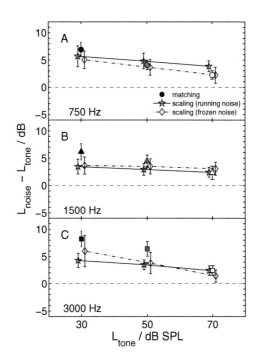

Figure 4.5: Mean results of Experiment 3. Level difference at equal loudness between noise and a pure tone as a function of the signal level for signals centered at 750 Hz (panel A), 1500 Hz (panel B) and 3000 Hz (panel C) using running noise (stars) or frozen noise (diamonds). For comparison corresponding results of Experiment 2 are shown (circles in panel A, triangles in panel B and squares in panel C). Error bars indicate interindividual standard errors. To increase readability data are slightly shifted against each other.

frequencies, the level differences measured with the categorical scaling procedure are in the range of the data gathered with the matching procedure. For each center frequency, a two-way analysis of variances (ANOVA) on the factors tone level and measurement method (three level: matching, scaling with running noise, scaling with frozen noise) showed no significant effect of the measurement procedure or the noise condition on the level difference at equal loudness (750 Hz: $F(2) = 0.410$, $p > 0.6$; 1500 Hz: $F(2) = 0.450$, $p > 0.6$; 3000 Hz: $F(1.240) = 0.796$, $p > 0.46$; Greenhouse Geisser used for 3000 Hz). A significant effect of the level was found for all center frequencies (750 Hz: $F(1.144) = 14.574$, $p \leq 0.05$; 1500 Hz: $F(1.192) = 5.237$, $p \leq 0.05$; 3000 Hz: $F(1.143) = 20.545$, $p \leq 0.01$; Greenhouse Geisser used for all center frequencies).

4.5 Discussion

4.5.1 Comparison to previous studies

For subcritical bandwidths it is commonly assumed that loudness is independent of bandwidth. This assumption is, among others, based on the extensive study on the influence of bandwidth on loudness by Zwicker *et al.* (1957). They approximated their results for loudness of subcritical bandpass noises with a horizontal line (see their Figs. 8-10). However, at some levels, their data show a slight decrease in loudness with bandwidth for subcritical bandwidths, which is in agreement with the present data. Examples are the results at 50 dB for a 210-Hz reference bandwidth and a center frequency of 1420 Hz, or at 60 dB for a 440-Hz center frequency, where the reference was a pure tone. Other studies also measured a higher loudness for a tone than an equal-level noise. For example, a positive level difference between a critical band wide noise and an equally loud pure tone was measured by Zwicker (1974) for center frequencies of 1, 1.5, and 2 kHz and levels of 30, 50, 70, and 90 dB. As in the present study, the level difference at equal loudness decreased with reference level. For 1.5 kHz, Zwicker measured a decrease of the level difference between a Bark wide noise and a pure tone at equal loudness by 5 dB when the reference level was increased from 30 dB to 70 dB. This is very similar to the 4 dB found in the present study. A level difference of 1.5 dB between narrowband noise and an equally loud pure tone was also reported by Grimm *et al.* (2002) for a 32-Hz wide multiplied noise centered at 2000 Hz and a reference level of 65 dB. This is similar to the 2 dB measured in the first experiment of the present study for a center frequency 3000 Hz and a bandwidth of 45 Hz. A 2-dB difference was also found by Florentine *et al.* (1978) for a 1-kHz pure tone and an equally loud 220-Hz wide noise.

The present data also agree with the data shown in Chapter 3 for a center frequency of 1500 Hz and a reference level of 50 dB. The maximum effect (measured at 135-Hz bandwidth) is about the same in these data. In Chapter 3 it was shown that the effect can be measured for noise and tone references, i.e., it does not depend on the choice of the reference. The present study extended this result showing that it is also observed using a different procedure. There is a tendency of a slightly smaller effect at the lowest level (30 dB) when categorical scaling is used instead of matching. A similar result was found in Anweiler and Verhey (2006) at their lower level (45 dB). However, the effect in the present study was not significant. This is in agreement with a recent study on loudness of speech signals where the data of categorical scaling and matching did not differ significantly from each other either (Rennies *et al.*, 2013). In contrast to the present study, Valente *et al.* (2011) observed that temporal integration in loudness was slightly smaller when measured using a matching procedure than when derived from scaling data. Thus, scaling does not always lead to smaller effects. Overall, they concluded that scaling is a good tool to measure levels at equal loudness. The present study supports this hypothesis.

4.5.2 Role of envelope fluctuations

Noise has a stochastic envelope which is different for each realization of the noise. However, this stochasticity is apparently not the reason for the difference in loudness between a tone and a noise, since random noise and frozen noise yield to similar results in Experiment 3 (Sec. 4.4). For the Gaussian noise used here, the spectrum of the envelope fluctuations is triangularly shaped with a maximum rate equal to the bandwidth of the noise (Lawson and Uhlenbeck, 1950; Dau *et al.*, 1999), i.e., the average fluctuation rate increases as the bandwidth increases[2]. Zhang and Zeng (1997) showed that the loudness of amplitude-modulated 1-kHz pure tones decreases with modulation rate for modulation frequencies below about 100 Hz, i.e., when the overall bandwidth of the modulated sound is still within the excitatory region of the critical band centered at the carrier frequency (see also Bauch, 1956). A similar effect is found in the present data, where the loudness decreases as the average envelope fluctuation rate increases for subcritical bandwidths. However, the data of Zhang and Zeng (1997) indicate that, at low modulation rates loudness is higher for a modulated than an unmodulated pure tone. In contrast, the present data indicate that a pure tone and an equal-level narrowband noise with a bandwidth of 15 Hz have about the same loudness. This finding is in agreement with Moore *et al.* (1999) and Grimm *et al.* (2002), who measured hardly any difference in level ($< 1\,\mathrm{dB}$) for equally loud modulated and unmodulated pure tones. None of the loudness studies using sinusoidally amplitude-modulated

---

[2]Other noise types may have different envelope spectra and average envelope rates (see, e.g., Dau *et al.*, 1999)

pure tones found effects of modulations larger than 3 dB, when the signal spectrum was still within the excitatory region of the critical band, whereas the effect of the present study was up to 8 dB. Thus, one may argue that the influence of envelope fluctuations on loudness is not sufficiently large to account for loudness of noise with a subcritical bandwidth.

Indirect evidence of a fundamental differences between the loudness of ampli-tude-modulated pure tones and narrow bands of noise were given in two recent studies by Soeta and colleagues (Soeta *et al.*, 2005; Soeta and Nakagawa, 2006). Using magnetoencephalography (MEG), they measured the response of the auditory system to sinusoidally amplitude-modulated tones (Soeta and Nakagawa, 2006). The characteristics of the negativity at about 100 ms after stimulus onset (N1m) are thought to represent the loudness of the sound (e.g., Morita *et al.*, 2003). The N1m amplitude hardly changed with modulation frequency for small frequencies, whereas N1m increased with frequency as soon as the sidebands fell outside the critical band around the carrier frequency. This latter effect was attributed to spectral loudness summation. In contrast, for bandpass noise with a subcritical bandwidth, N1m amplitudes decreased as the bandwidth increased. This latter effect is in agreement with the findings of the present study that loudness decreases with bandwidth for sounds with a subcritical bandwidth.

4.5.3 Role of spectral characteristics

If spectrum determines the effect, similar results should be obtained for the three center frequencies when replotted on a scale reflecting the frequency selectivity of the auditory system. Figure 4.6 shows, as an example, the data plotted on the Bark scale. The results are similar when the data are plotted on an ERB$_N$ scale (not shown). Panel A shows the data of the second experiment for the reference level of 30 dB, panel B those for the

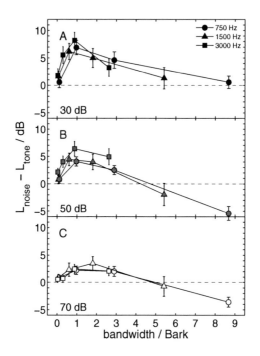

Figure 4.6: Mean results of Experiment 2. The data shown in Fig 4.3 are rearranged, now showing the level difference at equal loudness as a function of the bandwidth in Bark. Each panel shows the results for one reference level: 30 dB (panel A, black filled symbols), 50 dB (panel B, gray filled symbols) and 70 dB (panel C, open symbols). As in Fig 4.3 the symbols indicate the center frequency: 750 Hz (circles), 1500 Hz (triangles) and 3000 Hz (squares). Error bars indicate interindividual standard errors.

50 dB, and panel C the data for 70 dB. The symbols are the same as used in Fig 4.3. Using this scale the data of the three center frequencies are very similar in shape. There is a tendency of a shift of the maximum of the effect towards higher bandwidths at the highest level. This may reflect the increase in auditory filter width with level (Lutfi and Patterson, 1984). Thus, the effect of bandwidth on loudness may be due to the spectral analysis of the auditory system. However, a positive level difference as found in the present data cannot be understood on the basis of auditory frequency selectivity alone. Instead, a nonlinear process is required.

In Chapter 3 it was hypothesized that the effect may be due to suppression. Physiological studies at the level of the auditory nerve indicate that the suppressive regions can be found above and below the best center frequency of the fibres (Harris, 1979; Sellick and Russel, 1979; Schmiedt, 1982). The excitatory and inhibitory regions partly overlap. A larger overlap is found at higher stages of the auditory system (e.g., Sutter and Loftus, 2003). Thus, it is possible that suppression already occurs within the auditory filter at the signal frequency. Forward masking data support the notion that auditory filters derived from simultaneous masking data may be 1.2 to 1.8 times broader than the actual excitatory filters (Buchholz *et al.*, 2010). The data of the second experiment and those of Chapter 3 indicate that the effect starts at very narrow bandwidths, as narrow as 45 Hz at a center frequency of 1500 Hz. The suppressive regions measured in psychoacoustics were as close as 1.15 times the center frequency (Shannon, 1976; Ernst *et al.*, 2010). The data of Ernst *et al.* (2010) indicate that suppression may be observed when suppressor and suppressee have the same level, but the effect was not significant. It remains to be seen if suppression is the underlying process or if there are other mechanisms underlying the decrease in loudness with bandwidth for subcritical bandwidths. An alternative mechanism may be based on perceptual differences of the sound. The pitch strength decreases continuously as the bandwidth increases (Fastl and Zwicker, 2007, see their Fig. 5.29). This decrease is already observed at a bandwidth of 31.6 Hz and is similar for the range of frequencies used in the present study. Assuming that the auditory system assigns a higher loudness to tonal signals than to noise-like signals, the decrease in loudness with bandwidth may reflect the reduction in tonal character when the bandwidth is increased.

4.6 Summary and conclusion

Loudness of subcritical signals was measured for various center frequencies and levels. For all combinations of level and center frequency, the results consistently showed (i) hardly any difference in level between a pure and an equally-loud noise with a very narrow bandwidth, (ii) a positive level difference at intermediate bandwidths with a maximum up to 8 dB close to the critical bandwidth, and (iii) a decrease of the level difference as the bandwidth increases for supercritical bandwidths. The effect was observed using matching and scaling procedures. Thus, the dependency of loudness on the bandwidth for subcritical sounds seems to be a universal property of the auditory system. The underlying mechanism is still unclear but is presumably due to spectral properties rather than due to the inherent envelope fluctuations (i.e., temporal aspects) of the sound. The effect may be due to suppression or another more central process which associates a higher loudness to signals with a high pitch strength. Further studies are required to decide which is the underlying process.

5 Loudness of sounds with a subcritical bandwidth in hearing-impaired listeners

5.1 Introduction

The psychoacoustic experiments described in Chapter 3 and Chapter 4 showed that narrowband noise bands have a higher level than an equally loud pure tone at the center frequency of the noise. The maximum of this effect (up to about 8 dB) is observed for bandwidths close to the critical bandwidth. It decreases towards narrower bandwidths down to a level difference of about 0 dB at bandwidths of only a few Hertz. For supercritical bandwidths, a further increase in bandwidth leads to a decrease of the effect and the level difference becomes negative for broad bandwidths due to spectral loudness summation. The data presented in Chapter 4 showed that the effect was also level dependent. It was largest at low levels. These findings are difficult to reconcile with the current understanding of loudness. Models based on the long-term spectrum of the sound predict no level difference between subcritical noise bands and tones (see, e.g., Fastl and Zwicker, 2007). Models considering the temporal properties of the signal predict a lower level of the noise compared to an equally loud tone, i.e., the opposite effect (see Chapter 3). The mechanism underlying the found effect is still unclear. Cochlear compression and inherent level fluctuations of the noise are unlikely candidates to explain the effect. Other nonlinear mechanisms of the auditory system such as suppression may result in positive level differences between noise and pure tone at equal loudness as observed in the data (see Chapter 3 and Chapter 4). This would be a peripheral process, i.e., at the cochlear level. Alternatively the effect could be due to a more central process such as a difference of perception of tones and noise due to different pitch strengths (see Chapter 4).

The aim of this chapter is to investigate whether the underlying mechanism is either a peripheral or a more central process of the auditory system. To this end, level differences at equal loudness are determined in listeners with a cochlear hearing impairment. In this group of listeners suppression is reduced (see, e.g., Ernst *et al.*, 2010). In addition, the critical band filters are broader than in normal-hearing listeners (see, e.g., Nitschmann *et al.*, 2010). Thus, if the effect is due to a peripheral process, it should be reduced due to a reduced suppression and the broadening of the filters should be reflected in the position of the maximum of the effect. If in contrast it is due to a more central process, the magnitude of the effect should be similar to that found in normal-hearing listeners at least for subcritical bandwidths. Spectral loudness summation, observed at large bandwidths, is assumed to be due to critical band filtering and compression. Thus, it is reduced in hearing-impaired listeners (see, e.g., Verhey *et al.*, 2006; Brand and Hohmann, 2001). This reduction should also affect the level difference at equal loudness for bandwidths being slightly larger than the critical bandwidth.

5.2 Methods

Levels at equal loudness were measured using an adaptive loudness matching procedure. The procedure, stimuli generation, and presentation are the same as used for the normal hearing participants in Chapter 3 and Chapter 4. The experiment was controlled using the MATLAB AFC framework described by Ewert (2013). The stimuli were generated at a sampling frequency of 44.1 kHz. They were D/A converted and presented monaurally via an external sound card (RME Fireface 400) and headphones (Sennheiser HD650). The participants were seated in a sound-proof booth. The experimental setup was identical to the one used in Experiment 2 of Chapter 4 (see Sec. 4.3.1).

The test stimuli were Gaussian noises geometrically centered at 1500 Hz with bandwidths of 15, 45, 135, 405, 810, 1215, or 1620 Hz. The reference signal was a 1.5-kHz pure tone. For the normal-hearing listeners, the measured effect was particularly large at a reference level of 30 dB (see Chapter 4). To measure at a loudness comparable to the loudness of a 30-dB reference level in normal-hearing listeners, the reference level was chosen individually for each hearing impaired listener on the basis of the following equation:

$$thr_{HI} + \left(\frac{ucl_{HI} - thr_{HI}}{ucl_{NH} - thr_{NH}} \right) \cdot 30dB \qquad (5.1)$$

where thr_{HI} is the individual threshold in quiet of the participant at 1.5 kHz, ucl_{HI} and ucl_{NH} are the uncomfortable levels for hearing-impaired (HI) and normal-hearing (NH) listeners and thr_{NH} is the threshold in quiet at 1.5 kHz for a normal-hearing listener. thr_{NH} was set to 0 dB. The levels at uncomfortable loudness ucl_{HI} and ucl_{NH} were taken from Keller (2006), who showed that the uncomfortable levels for normal-hearing and hearing-impaired listeners were similar ranging from 100 to 110 dB. Thus, in this study, ucl_{HI} as well as ucl_{NH} were set to a value of 100 dB. Prior to the matching experiment, the 1.5-kHz pure tone was presented to the participants at the calculated reference level. The participants were asked if the stimulus had a comfortable loudness. If it was too soft or too loud the reference level was shifted by 6 dB towards the desired direction. This was necessary for one participant: For participant H6, the level had to be increased by 6 dB. Reference and test stimuli were 500 ms long including 5-ms, raised cosine ramps at stimulus on- and offset. Test and reference signal were matched in loudness using an adaptive two-interval, two-alternative forced-choice procedure. For each bandwidth three starting levels were used: -10 dB, 0 dB, and 10 dB relative to the reference level. The tracks for all bandwidths and starting levels were interleaved. In each trial, a test and a reference stimulus were presented in random order, separated by a silent interval of 500 ms. The participants indicated which interval they perceived as the louder one by pressing the corresponding key. The level of the test stimulus was varied according to a one-up one-down rule. The maximum allowed level of the test stimulus was 90 dB. The step size at the beginning of a track was 8 dB. It was halved after every upper reversal until a step size of 2 dB was reached. With this minimum step size the track continued for four reversals. To estimate the level at equal loudness, the mean of these four last reversals was calculated. The final estimate was taken as the average of the results for the three starting levels.

Nine hearing-impaired listeners (5 male, 4 female) aged from 22 to 71 years participated in the experiment. Criteria to participate in the experiment were (i) that they had a flat hearing loss between 40 and 75 dB HL around the frequency of 1.5 kHz and (ii) that they did not report a tinnitus in the investigated frequency range. For the experiment, the ear which better fulfilled the above criteria was selected. For all participants the threshold of the other ear did not allow a cross talking of the stimulus. The individual audiograms are shown in Fig. 8.11 in the Appendix 8.3. The gray area indicates the investigated frequency range. To determine the participants' hearing range at the frequency of 1.5 kHz, a categorical loudness scaling for a 1.5-kHz pure tone and for Gaussian noise stimuli centered at 1.5 kHz with a bandwidth of 135 Hz was conducted in the same way as in Experiment 3 in Chapter 4 (see Sec. 4.4.1). The resulting loudness functions for the noise stimuli are shown as solid black lines in Fig. 8.12 in the Appendix 8.3. For comparison, gray dashed lines show the averaged loudness function for the same type of stimuli of the normal-hearing participants in Experiment 3 in Chapter 4. All participants were paid volunteers.

5.3 Results

Figure 5.1 shows individual results of the nine participants (H1 to H9). Each panel shows the mean level difference between noise and pure tone at equal loudness of the three repetitions of the experiment as a function of bandwidth for one participant. Error bars indicate the intraindividual standard deviation. The standard deviation was usually less than 2 dB. The average standard deviation was about 1.5 dB. Vertical lines indicate the critical bandwidth at a frequency of 1500 Hz in equivalent rectangular bandwidth (ERB$_N$, Moore and Glasberg, 1981, dotted line) and in Bark (Fastl and Zwicker, 2007, dash-dotted line). For participant H1, the level differences at equal loudness deviated from 0 dB by less than about 2 dB for all bandwidths from 15 to 1620 Hz. Participant H2 measured a level difference of -14 dB in the track for the starting level of -10 dB of the bandwidth of 1215 Hz. This level was below the individual threshold of this participant. After the experiment, the participant reported an inability to do the task properly in this particular track, because only one stimulus was audible. For this reason this track was excluded from the further analysis. In the other two tracks for this bandwidth the estimated

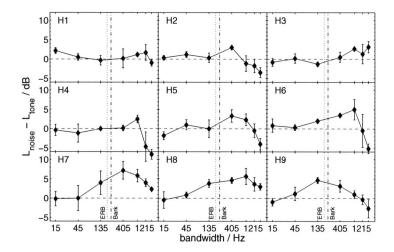

Figure 5.1: Individual results of the nine hearing-impaired participants. Level difference at equal loudness between noise and a pure tone as a function of the bandwidth of the noise. Error bars indicate intraindividual standard deviations. In each panel vertical lines show the critical bandwidth at a frequency of 1500 Hz in ERB_N (dotted line) and Bark (dash-dotted line).

level difference at equal loudness was about -1 and -3 dB, i.e., the adaptive procedure converged to a clearly audible level. For the bandwidths from 15 to 135 Hz the level difference for this participant was close to 0 dB. At the bandwidth of 405 Hz it reached a value of 3 dB. For the larger bandwidths it decreased to about -4 dB at 1620 Hz. The participants H3 and H4 showed a level difference close to 0 dB for bandwidths up to 405 Hz. For the bandwidth of 810 Hz the level difference increased to about 3 dB for both participants. For H3, a level difference of about 3 dB was also found for the bandwidths of 1215 and 1620 Hz. For H4, the level differences decreased to about -5 dB at 1215 Hz and -7 dB at 1620 Hz. For the participants H5 to H8, the measured maximum level difference had larger values between about 3 dB (H5) and 7 dB (H7), which was found at a bandwidth of 405 Hz (H5, H7) or 1215 Hz (H6, H8). For H6, H7 and H8, it had values close to 0 dB for the bandwidths from 15 to 45 Hz. For H5 it deviated from 0 dB by less than 2 dB for the bandwidths from 15 to 135 Hz. For larger bandwidths, these participants showed a decrease in the level difference. At the largest bandwidth of 1620 Hz, it still had a positive level between about 2 to 3 dB for H7 and H8, whereas a negative level difference of about -4 to -5 dB was measured at this bandwidth for H5 and H6. For participant H9 the maximum level difference of about 5 dB was found at a bandwidth of 135 Hz. For the narrower bandwidths it was close to 0 dB. For larger bandwidths it slowly decreased to about -3 dB at the bandwidth of 1620 Hz.

The mean data across all participants are shown in Fig. 5.2. As in Fig. 5.1, the level difference at equal loudness is shown as a function of noise bandwidth. Error bars indicate the interindividual standard error. Vertical lines again indicate the critical bandwidth at the frequency of 1500 Hz in ERB_N (dotted line) and Bark (dash-dotted line). The level difference at equal loudness is close to 0 dB for the two narrowest noise signals. As the bandwidth increases, the level difference at equal loudness

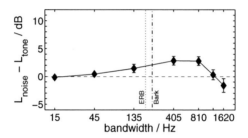

Figure 5.2: Mean results of the nine hearing-impaired participants. Level difference at equal loudness between noise and a pure tone as a function of the bandwidth of the noise. Error bars indicate interindividual standard errors. Vertical lines show the critical bandwidth at a frequency of 1500 Hz in ERB_N (dotted line) and Bark (dash-dotted line).

increases up to the bandwidth of 810 Hz where it is about 3 dB. For broader noises, it decreases with increasing bandwidth. At the bandwidth of 1620 Hz, it amounts to -1.5 dB.

5.4 Discussion

Figure 5.3 shows the mean data of the hearing-impaired participants as shown in Fig. 5.2 (black diamonds). In addition the mean data of the 10 normal-hearing participants of Experiment 2 in Chapter 4 for the center frequency of 1500 Hz and reference levels of 30 dB (filled gray triangles) as well as 70 dB (open gray triangles) are shown for comparison. With the reference level calculated by equation (5.1) the loudness for the hearing-impaired listeners is comparable to that of the reference level of 30 dB for the normal-hearing listeners[1]. Compared to the normal-hearing data for the reference level of 30 dB (filled gray triangles in Fig. 5.3), the maximum level difference at equal loudness is markedly reduced (about 6 dB for normal-hearing and 3 dB for hearing-impaired) and shifted towards a broader bandwidth (135 Hz for normal-hearing and 810 Hz for hearing-impaired). The decrease in level difference towards narrower bandwidths is shallower than in the normal-hearing listeners. The shifting of the maximum may be explained by the broadening of the auditory filters (Nitschmann *et al.*, 2010; Baker and Rosen, 2002). For bandwidths broader than the bandwidth with the maximum level difference the decrease seems to be steeper in the hearing-impaired than in the normal-hearing listeners. Due to the reduced spectral loudness summation in hearing-impaired listeners (Verhey *et al.*, 2006; Brand and Hohmann, 2001) the opposite effect would be expected. This effect is presumably ob-

---

[1] The calculated reference level does not lead to the same sensation level, i.e., level above threshold. The mean sensation level for the hearing-impaired listeners is only about 16 dB SL.

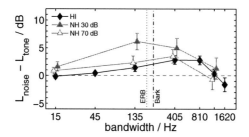

Figure 5.3: Comparison of hearing-impaired and normal-hearing data. Mean level differences at equal loudness between noise and a pure tone are shown as a function of the bandwidth of the noise for the nine hearing-impaired participants as already shown in Fig. 5.2 (black diamonds) and for the 10 normal-hearing participants of Experiment 3 in Chapter 4 for a reference level of 30 dB (gray filled triangles) and 70 dB (gray open triangles) as already shown in panel B of Fig. 4.3 in Chapter 4. Error bars indicate interindividual standard errors. Vertical lines show the critical bandwidth at a frequency of 1500 Hz in ERB_N (dotted line) and Bark (dash-dotted line). To increase readability data are slightly shifted against each other.

served for broader bandwidths that were not tested here. Based on this comparison at approximately the same loudness one may conclude that a peripheral nonlinearity causes the found effect, since the maximum level difference is reduced.

An alternative is the comparison at the same level. The reference level used for the hearing-impaired listeners is close to the reference level of 70 dB of the normal-hearing listeners. The mean reference level for the hearing-impaired listeners is 67.5 dB. Compared to these data (open gray triangles in Fig. 5.3) the maximum level difference deviates by only about 0.5 dB between the two groups of listeners. For the hearing-impaired listeners it again seems to be shifted to a broader bandwidth. However, the bandwidth resolution is reduced in the normal-hearing data (810 and 1620 Hz were not measured). Therefore a conclusive assessment can not be made at this point. The two data sets seem to be very similar for the bandwidths measured in both groups. Thus, from this comparison one may conclude that the hearing-impairment does not change the effect.

Previous studies compared the results of normal-hearing and hearing-impaired listeners at both, the same level and approximately the same loudness (see, e.g., Verhey *et al.*, 2006; Nitschmann *et al.*, 2010). Some data in the literature indicate that a comparison at the same level above threshold leads to more comparable results between hearing-impaired and normal-hearing listeners (see, e.g., Glasberg *et al.*, 1987; Baker and Rosen, 2002). On this basis one has to conclude that a peripheral nonlinearity is the underlying mechanism of the effect.

The small difference at very narrow bandwidths between the two groups indicates that it is presumably not the difference in slope of the loudness function, commonly associated with a difference in compression. The reduced compression leads to a higher perceived modulation depth in the impaired ear (see

Moore *et al.*, 1996). Thus, the influence of the inherent level fluctuations in the comparison between tones and noise should be higher than for the normal-hearing listeners, resulting in a higher level difference for the hearing-impaired listeners, if the effect is based on level fluctuations. However, this can not be seen in the data.

In Chapter 4 the perception of pitch strength as a more central process was discussed as the underlying mechanism of the found effect. This was motivated by the effect of bandwidth on pitch strength, which was comparable to the effect of bandwidth on loudness found in Chapter 4. The pitch strength of pure tones is higher than that of narrowband noise (see Hansen *et al.*, 2011). Pitch strength decreases continuously as the bandwidth increases (Fastl and Zwicker, 2007, see their Fig. 5.29). This approach assumes that a tonal signal is more intrusive and thus louder than a noise signal, i.e., a signal with a large pitch strength is louder than one with a smaller pitch strength. Although the perception of pitch strength is presumably a high level process it may be modulated by lower level processes in some cases. For example, Leek and Summers (2001) found that the perception of pitch strength for iterated rippled noise is reduced in listeners with cochlear damage. They suggested that this derives from a poorer frequency resolution of the auditory periphery and an associated deficit in temporal coding. However, one should keep in mind that pitch strength strongly depends on the stimulus type (see Hansen *et al.*, 2011, for a review). Thus, results for iterated rippled noise may not be directly comparable to the pitch strength for narrowband noises. Currently it is unclear if the pitch strength is reduced by the similar amount as found in the present loudness data.

A potential peripheral process accounting for the effect of bandwidth on loudness is suppression as mentioned in Chapter 4. Ernst *et al.* (2010) showed that suppression is reduced in hearing-impaired listeners. If suppression is the underlying mechanism, one would expect a reduced effect of bandwidth on loudness as found in the data, at least if compared at the same reference loudness for hearing-impaired and normal-hearing participants.

In summary, although a peripheral process is a likely candidate, one can not rule out that higher processes also contribute to the effect of a positive level difference between a narrowband noise and an equally loud tone. Further studies, such as the determination of suppression and the perception of pitch strength in the participants of this study are required to quantify the contribution of low and high level processes.

5.5 Summary and conclusion

This study measured the effect of bandwidth on loudness in hearing-impaired participants with a special focus on subcritical bandwidths. For these subcritical bandwidths normal-hearing participants showed a positive level difference between a noise and an equally loud tone. The data of the hearing-impaired participants also showed a positive level difference. It was reduced compared to the data of the normal-hearing participants and the maximum was shifted towards broader bandwidths. This latter effect is presumably due to a reduced frequency selectivity. Overall the data indicate that the underlying mechanism is mainly peripheral rather than central. In combination with the data of normal-hearing participants, the data of this chapter can serve as a basis for developing a revised loudness model.

6 Increased intensity discrimination thresholds in patients suffering from tinnitus with a normal audiogram[1]

6.1 Introduction

Normal hearing thresholds have long been regarded as an indicator for the absence of cochlear damage. However, it has recently been shown in mice that sound exposure at nightclub noise levels, which leads to a temporary shift of the hearing thresholds, also causes permanent deafferentation of a large fraction of the auditory nerve (AN) fibres in the high-frequency range (Kujawa and Liberman, 2009). The effect of this cochlear damage also manifested in a marked reduction of the amplitude of the wave I potential (generated through the activation of AN

---

[1] A modified Version of this chapter is published as:

Epp, B., Hots, J., Verhey, J.L. and Schaette, R. **(2012)** "Increased intensity discrimination thresholds in patients suffering from tinnitus with a normal audiogram" J. Acoust. Soc. Am. **132** (3) EL196–EL201.

fibres) of the auditory brainstem response (ABR) at high sound intensities, whereas responses at low sound intensities were almost unchanged, suggesting that the deafferentation predominantly affected high-threshold fibres. A similar reduction of ABR-wave I at high sound intensities has also been observed in subjects with tinnitus and a normal audiogram (Schaette and McAlpine, 2011). Moreover, normal-hearing tinnitus subjects also showed increased tone-detection thresholds in threshold-equalizing noise at high intensities, which has been interpreted as evidence for central deafferentation despite normal-hearing thresholds (Weisz *et al.*, 2006).

The wide dynamic range of sounds in the acoustic environment is covered by the different types of AN fibres: the dynamic range of the response of high spontaneous rate fibres covers low sound intensities, whereas fibres with medium and low spontaneous firing rates are sensitive to higher sound intensities (Liberman and Kiang, 1978; Yates *et al.*, 1990). Deafferentation of AN fibres might thus impair the ability to discriminate sound intensities, as deafferentation reduces the number of fibres that are available to resolve intensity differences. Specifically, when high-threshold fibres are deafferented, as it seems to be the case in noise-exposed mice, one would expect to see a deficit at medium to high sound intensities.

If tinnitus patients with normal audiograms do have deafferentation of AN fibres, as suggested by ABR measurements (Schaette and McAlpine, 2011) and hearing tests in background noise (Weisz *et al.*, 2006), they should thus also show an impairment in an intensity discrimination task. Moreover, if the deafferentation is linked to the development of tinnitus, as indicated by computational modeling (Schaette and McAlpine, 2011), the effect should be most pronounced in the tinnitus frequency range. To test this hypothesis, just-noticeable intensity differences (intensity JNDs) for pure tones were measured in normal-hearing listeners reporting tinnitus, and in control participants with normal hearing. In a first experiment, pure-tone detection thresholds were measured at the relevant probe frequencies, including a frequency in the peak region of the average tinnitus spectrum of the participants. A bandpass noise was simultaneously presented to reduce off-frequency listening in adjacent auditory filters. In a second experiment, intensity JNDs were measured for the same frequencies in the presence of the same maskers.

6.2 Methods

For each listener, hearing thresholds in quiet were measured using a calibrated clinical audiometer for frequencies of 125 to 8000 Hz. Subsequently, masked thresholds and pure tone intensity JNDs at frequencies of 1000, 2450, and 6000 Hz were assessed. Measurements were done monaurally for both ears of each listener. For both measures, an adaptive three-alternative forced-choice procedure with a one-up two-down rule was used, resulting in the 70.7% point of the psychometric function (Levitt, 1971). Signals were generated in MATLAB with a sampling rate of 96000 Hz and 32-bit resolution. Signals were presented via headphones (Sennheiser HDA200) using an external sound card (RME Fireface 400) in a sound-proof booth. Each interval had a duration of 500 ms and the intervals were separated by 500 ms of silence. Visual feedback was given after each trial. Off-frequency listening was reduced by presenting two additional one-half octave wide noise bands with a total level of 50 dB SPL above and below the signal of the tone. The noise bands were generated by bandpass-filtering a white noise using a 8th-order Butterworth bandpass filter at center frequencies one octave above and below the target signal frequency.

6.2.1 Masked thresholds

Three intervals were presented to the listener, each containing an independent realization of the masker. One randomly chosen interval also contained the target signal. The listener had to indicate which of the intervals contained the target signal. The initial level of the target signal was 8.5 dB SPL. The step size of the adaptive procedure was initially 4 dB and was halved after each upper reversal of the tracking procedure until the smallest step size of 1 dB was reached. The arithmetic mean of six reversals using the smallest step size was used to estimate the threshold. Individual thresholds were calculated as the arithmetic mean of three repetitions.

6.2.2 Intensity JND

Three intervals were presented to the listener, each containing an independent realization of the masker and the tone at a fixed level (in the following referred to as reference). The reference tone had levels of 30, 50 or 70 dB SPL (in the following referred to as reference level). In one randomly chosen interval another tone with the same frequency was added in phase to the reference tone, i.e., the amplitude of the resulting tone was equal to the sum of the amplitudes of the reference and the added tone. The listener had to indicate in which interval the intensity of the tone was higher. The step size of the intensity increment calculated as $\frac{\Delta I}{I}$ was initially set to 0.3 and was halved after each upper reversal until the smallest increment of 0.0375 was reached. The arithmetic mean of six reversals using the smallest step size was used to estimate the JND. Individual JNDs were calculated as the arithmetic mean of three repetitions.

6.2.3 Tinnitus spectrum measurements

Tinnitus pitch was characterized using a modified version of the tinnitus spectrum approach (Norena *et al.*, 2002). Comparison sounds (pure tones of 250, 500, 1000, 1500, 2000, 3000, 4000, 6000, 8000, 12000, and 16000 Hz) were generated using custom-made MATLAB software and were presented via Sennheiser HD600 headphones. The comparison tones were first matched to the loudness of the tinnitus using a single-interval adaptive procedure (Lecluyse and Meddis, 2009). During the loudness matching, the sound intensity was limited to levels $\leq 100\,\mathrm{dB\,SPL}$. To obtain tinnitus pitch similarity ratings, the loudness-matched tones were presented in random order, and following each presentation, participants were asked to rate the similarity between the pitch of the comparison tone and their tinnitus on a scale from 0 to 10, with 0 for "completely different" and 10 for "extremely similar". Each tone was presented and rated three times. For comparison tones of 16 kHz, a loudness match could not be achieved for all participants, and thus the results for this frequency were excluded from further analysis.

6.2.4 Listeners

Fourteen listeners without tinnitus (control group, mean age 26 ± 2 years, 10 female) and 11 listeners who reported a tinnitus (tinnitus group, mean age 38 ± 4 years, all female) participated in the experiment. All listeners had pure-tone thresholds in quiet $\leq 20\,\mathrm{dB\,HL}$ in the frequency range from 250 Hz to 8 kHz. One of the listeners of the tinnitus group reported a tinnitus in the left ear only, and there were two listeners with tinnitus dominant in the left ear and two listeners with tinnitus dominant in the right ear. The remaining tinnitus subjects reported a tinnitus in either both ears or located within the head. For the participant with unilateral tinnitus, only the tinnitus ear was included in the analysis.

6.2.5 Statistical analysis

To test for significant differences, the Mann-Whitney U-Test was employed. For audiometric and masked thresholds, two-sided tests were used. For comparing intensity JNDs between the tinnitus and the control group, one-sided tests were used, since it was expected to see higher intensity JNDs in the tinnitus group. Differences were considered to be significant for $p < 0.05$. All data analysis was carried out using MATLAB.

6.3 Results and Discussion

The mean audiograms of the listener groups are shown in Fig. 6.1a (gray: control; black: tinnitus). There were no significant differences in hearing thresholds of the two groups at any frequency. Tinnitus spectrum measurements showed that pure tones of 6 and 8 kHz were rated as most similar to the tinnitus pitch (Fig. 6.1b).

Masked thresholds and intensity JNDs at 1000, 2450, and 6000 Hz were measured (see Sec. 6.2). The highest frequency was chosen to be in the peak region of the average tinnitus spectrum (Fig. 6.1b). The average masked thresholds and intensity JNDs for control ($n = 28$ ears) and tinnitus participants ($n = 21$ ears) are shown in Figure 6.2 (gray lines: control; black lines: tinnitus).

For signal frequencies of 1000 and 2450 Hz, the intensity JNDs were rather similar for reference levels of 30 and 50 dB SPL, and showed a tendency to decrease for a reference level of 70 dB SPL. This pattern was observed in both the tinnitus and the control group. At 1000 Hz, there were no significant differences in JNDs between the control and the tinnitus group at all levels. At 2450 Hz, the average JND of the tinnitus group was significantly higher for a reference level of 70 dB SPL. For a signal frequency of 6 kHz, the pattern of intensity JNDs vs. sound

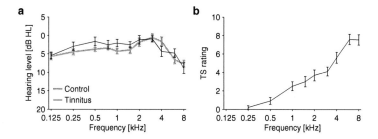

Figure 6.1: Audiograms and tinnitus spectra. **a)** Mean audiograms of all tinnitus (black) and control participants (gray). **b)** Average tinnitus spectrum of the tinnitus participants, error bars indicate ±1 standard error of the mean. Tinnitus pitch similarity ratings, on a scale from 0 to 10 with 0 for "completely different" and 10 for "extremely similar", were obtained using pure tone stimuli.

intensity differed between the tinnitus and the control group, with the tinnitus group displaying a more pronounced "mid-level hump". JNDs did not differ significantly at reference levels of 30 and 70 dB SPL, but were significantly higher in the tinnitus group at a reference level of 50 dB SPL.

The presence of increased intensity JNDs for 6000 Hz at 50 dB SPL but not at 70 dB SPL in the tinnitus group might seem at odds with the hypothesized deafferentation of high-threshold AN fibres, as one would expect this kind of cochlear damage to lead to intensity discrimination deficits at medium and high sound intensities. However, a comparison of the present results to ABR data from noise-exposed mice with histologically confirmed deafferentation of AN fibres (Kujawa and Liberman, 2009) shows a high similarity which offers a putative explanation: eight weeks after exposure to an octave band noise at 100 dB SPL, the ABR-wave I response of the mice showed a sub-

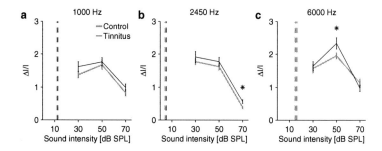

Figure 6.2: Masked thresholds (vertical lines) and intensity JNDs, grand mean over all control (gray) and tinnitus (black) participants. Results for target signal frequencies of 1000 Hz are shown in (a), 2450 Hz in (b), and 6000 Hz in (c). Intensity JNDs are shown as intensity increments $\Delta I/I$ as a function of the reference level. An asterisk indicates a statistical significant difference between control and tinnitus group ($p < 0.05$). Error bars indicate ± 1 standard error of the mean

stantial decrease in the slope of the amplitude growth function at levels around 40 to 60 dB SPL, whereas the slope was similar to control mice for higher sound intensities (Kujawa and Liberman, 2009, see also Fig. 6.3b of the present study). Assuming that the detection of an intensity increment is coded as an intensity-dependent increment in neural activity at the level of the auditory nerve, a shallower slope in the input-output (I/O) function would require a larger input increment to result in the same activity increment in the auditory nerve. The JNDs of tinnitus and control listeners differ for a reference level of 50 dB SPL, where the ABR I/O function of the noise-exposed mice is shallower (Fig. 6.3), while at 70 dB SPL, where the slope of the ABR I/O functions of noise-exposed and control mice is the same (Fig. 6.3), there is no significant difference between the control group and the tinnitus group. Therefore, our intensity JND results are generally consistent with electrophysiological data from mice with deafferentation of AN fibres.

Nine of the 11 tinnitus listeners also participated in a previous study by Schaette and McAlpine (2011), where a reduced wave I amplitude of the auditory brainstem response (ABR) was found in listeners with tinnitus and a normal audiogram. The deficit in intensity JNDs found in the present study in the tinnitus listeners might thus be linked to reduced ABR-wave I amplitudes, as found in mice with AN fibre deafferentation (Kujawa and Liberman, 2009). Taken together, these two results indicate that deafferentation of AN fibres could underlie tinnitus with a normal audiogram, which is further supported by the finding that listeners with tinnitus and normal hearing thresholds also show increased tone detection thresholds in high-level background noise in the tinnitus frequency range (Weisz *et al.*, 2006).

The two groups of subjects were matched with respect to their audiograms, but the medium age was higher in the tinnitus group than in the control group. He *et al.* (1998) found higher

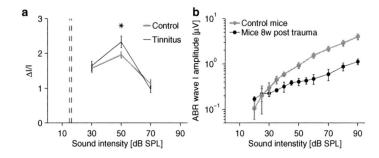

Figure 6.3: Comparison of psychoacoustical data with mouse auditory brainstem response (ABR) data. **a)** Intensity JNDs of tinnitus and control listeners for a frequency of 6000 Hz. **b)** ABR-wave I amplitude (tone pip stimuli, 32 kHz) of control mice (gray) and of noise-exposed mice eight weeks after exposure to octave-band noise (8-16 kHz) at 100 dB SPL for two hours (black), data courtesy of Kujawa and Liberman (2009). ABR-wave I, reflecting the summed activity of AN fibres, showed a particularly shallow growth around 50 dB SPL in the exposed mice.

intensity JNDs for low (40 dB) and high (80 dB) reference levels for older subjects (68-77 years) than for younger subjects (22-33 years). This may account for the (in most cases not significant) tendency of slightly higher JNDs in the tinnitus group for the 1000-Hz and 2450-Hz data. It is, however, unlikely that age effects account for 6000-Hz data, since (i) the effect of age in He *et al.* (1998) is considerably smaller for their highest frequency (4 kHz), and (ii) the JNDs of the present study are the same for both groups of subjects at the low and high reference levels and only significantly higher at 50 dB.

Apart from peripheral damage, also changes in central auditory processing could have influenced intensity discrimination thresholds in the tinnitus group. For example, sounds that are similar to the tinnitus might be processed by the same neuronal networks that generate the tinnitus, and thus they could also be affected by altered neuronal processing. Computational modeling studies have provided indications of how tinnitus might alter information processing in the central auditory system. For tinnitus after hearing loss (Schaette and Kempter, 2009) or AN fibre deafferentation with normal-hearing thresholds (Schaette and McAlpine, 2011), the models predict that plasticity trying to compensate for the resulting loss of AN signal will lead to an increase in neuronal response gain in the central auditory system. Increased gain in the tinnitus frequency range could also influence intensity discrimination. Future studies could try to address this in more detail by combining psychophysical and evoked potential measurements to study the relation between AN fibre deafferentation, psychophysical performance and neuronal responses in the central auditory system.

7 Summary and concluding remarks

This thesis investigated different aspects of the perception of loudness and sound intensity.

In Chapter 2 it was shown that an inclusion of compression as a nonlinear characteristic of the auditory system in a leaky integrator model leads to deviations of the predictions for temporal integration data at intermediate durations. The study by Rennies *et al.* (2010) showed similar deviations for dynamic loudness models. A new model approach for the prediction of temporal integration of loudness using two leaky integrators arranged in a parallel fashion was introduced in that chapter. The deviations between the predictions and experimental data were eliminated by this new model approach. Further, using the example of the *eDLM* it was shown, that the predictions of dynamic loudness models were improved for intermediate durations if the new temporal integration stage was included. The question if this improved prediction of temporal integration of loudness for intermediate durations results in an improved prediction of the effect of duration on spectral loudness summation for intermediate durations (see Sec. 1.2.4) could be an interesting issue of a future study.

The Chapters 3 to 5 focussed on the loudness perception of subcritical sounds. In Chapter 3 it was shown that, due to inherent level fluctuations, dynamic loudness models predicted a negative level difference between narrow bands of noise centered at 1.5 kHz and a 1.5-kHz pure tone at equal loudness. These predictions are not in agreement with the bandwidth independent predictions of loudness for subcritical sounds in stationary loudness models. Results of matching experiments between bands of noise and pure tones for different center frequencies and reference levels with normal-hearing participants in Chapter 3 and Chapter 4 are at odds with the predictions of both types of models. At equal loudness, a higher level for noise signals with a bandwidth close to the critical bandwidth than for a pure tone was found. For bandwidths as narrow as 5 or 15 Hz, this effect converged to a similar level for the pure tone and the noise. For bandwidths broader than the critical bandwidth it decreased with an increase of the bandwidth. The experiments in Chapter 3 showed that this effect is found for different reference conditions and the experiments in Chapter 4 revealed that the effect is largest at low levels and independent of the measurement procedure. The mechanism underlying the found effect is still unclear. Suppression and the perception of pitch strength were discussed as possible candidates to explain the findings. To test if the effect is due to a more central process (such as pitch strength) or a more peripheral process (such as suppression) in Chapter 5 a similar loudness matching experiment was conducted with hearing-impaired listeners. For a reference level at approximately the same loudness for normal-hearing and hearing-impaired listeners a reduced effect with a maximum shifted to broader bandwidths was found for the hearing-impaired participants. This indicated that the effect is mainly of a peripheral, rather than a central origin. To gain further insight on this topic experiments on the perception of pitch strength and suppression should be conducted with the hearing-impaired as well as the normal-hearing participants.

The comparison of intensity discrimination thresholds between normal-hearing participants and tinnitus participants with a normal audiogram in Chapter 6 showed that for the latter the ability to discriminate differences in the intensity was significantly reduced in the tinnitus frequency range at a reference level of 50 dB SPL. These findings are consistent with physiological data for auditory nerve fibre deafferentation by Kujawa and Liberman (2009). The results of this chapter substantiate the hypothesis that auditory nerve fibre deafferentation is a factor contributing to the emergence of tinnitus.

The data presented in this thesis provide new insights into the encoding of suprathreshold intensities in the auditory system, which are not covered by current models. These data can serve as a basis for developing a revised model, e.g., a revised loudness model for normal-hearing and hearing-impaired listeners.

8 Appendix

8.1 Individual results of the Experiment in Chapter 3

Figure 8.1 shows the individual results of the 10 participants described in Chapter 3. Each panel shows the level differences at equal loudness between a noise centered at 1.5 kHz and a 1.5-kHz pure tone as a function of the bandwidth of the noise. Blue circles indicate the results of the tone reference and red squares the results of the noise reference condition. Error bars indicate the intraindividual standard deviations of the three repetitions of the experiment. It was usually less than 2 dB. For the noise reference condition participant P7 had a larger standard deviation of up to about 6.5 dB. In general, the participants showed a similar effect in both conditions of the experiment, whereas participant P7 showed deviations between the conditions, which increased with the bandwidth. At the largest bandwidth of 405 Hz the level difference at equal loudness was about 14.5 dB larger for the noise reference condition. Participant P9 showed a larger level difference up to about 4.5 dB for the noise condition. Vertical lines show the critical bandwidth at a frequency of 1.5 kHz in ERB_N (dotted line) and Bark (dash-dotted line). To increase readability data are slightly shifted against each other.

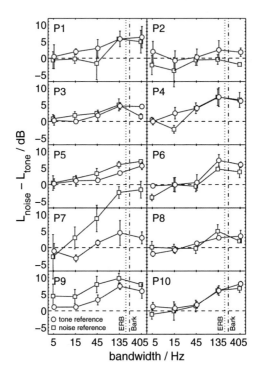

Figure 8.1: Individual results of the 10 participants in the experiment in Chapter 3. Level differences at equal loudness between a noise centered at 1.5 kHz and a 1.5-kHz pure tone as a function of the bandwidth of the noise for the tone reference (blue circles) and the noise reference (red squares) condition. Error bars indicate intraindividual standard deviations. Vertical lines show the critical bandwidth at a frequency of 1.5 kHz in ERB_N (dotted line) and Bark (dash-dotted line). To increase readability data are slightly shifted against each other.

8.2 Individual results of the Experiments in Chapter 4

Figure 8.2 shows individual level differences at equal loudness between noise centered at 3000 Hz and the 3000-Hz pure tone as a function of noise bandwidth for the nine participants in Experiment 1 in Chapter 4. Each panel shows individual data of one of the nine subjects. Error bars indicate intraindividual standard deviations of the three repetitions of the experiment. For all participants the standard deviation is rather small (usually less than 2 dB). Vertical lines indicate the critical bandwidth at a frequency of 3000 Hz in ERB_N (dotted line) and Bark (dash-dotted line). All participants measured a positive level difference at equal loudness at intermediate bandwidths with a maximum at 405, 810 or 1620 Hz. The size of the effect differed across subjects. It ranged from about 4 dB (P2 and P3) to 12 dB (P9).

Figures 8.3 to 8.5 show the individual results of Experiment 2 in Chapter 4 for the center frequencies of 750 Hz (Fig. 8.3), 1500 Hz (Fig. 8.4) and 3000 Hz (Fig. 8.5). Each panel shows the level difference at equal loudness between noise and pure tone as a function of noise bandwidth at a reference level of 30 dB SPL (black filled symbols), 50 dB SPL (gray filled symbols) and 70 dB SPL (open symbols) for one subject. Vertical lines again indicate the critical bandwidth at the corresponding frequency in ERB_N (dotted line) and Bark (dash-dotted line). To increase readability data are slightly shifted against each other.

Figures 8.6 to 8.10 show individual results of Experiment 3 in Chapter 4. Loudness functions for the pure tone (solid line) and the noise stimuli (dashed line) are shown for the center frequencies of 1500 Hz (Fig. 8.6) and 3000 Hz (Fig. 8.7) in the running noise condition as well as for the center frequencies of

750 Hz (Fig. 8.8) 1500 Hz (Fig. 8.9) and 3000 Hz (Fig. 8.10) in the frozen noise condition.

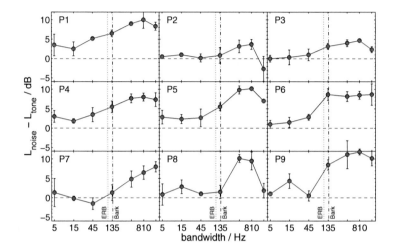

Figure 8.2: Individual results of the nine participants in Experiment 1 in Chapter 4 for signals centered at 3000 Hz. Level difference at equal loudness between noise and a pure tone as a function of the bandwidth of the noise at a reference level of 50 dB SPL. Error bars indicate intraindividual standard deviations. In each panel vertical lines show the critical bandwidth at a frequency of 3000 Hz in ERB_N (dotted line) and Bark (dash-dotted line).

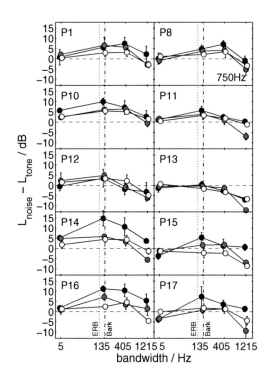

Figure 8.3: Individual results of the 10 participants in Experiment 2 in Chapter 4 for signals centered at 750 Hz. Level difference at equal loudness between noise and a pure tone as a function of the bandwidth of the noise at a reference level of 30 dB SPL (black filled symbols), 50 dB SPL (gray filled symbols) and 70 dB SPL (open symbols). Error bars indicate intraindividual standard deviations. In each panel vertical lines show the critical bandwidth at a frequency of 750 Hz in ERB_N (dotted line) and Bark (dash-dotted line). To increase readability data are slightly shifted against each other.

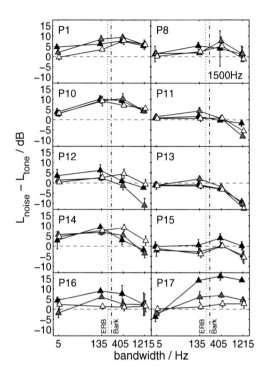

Figure 8.4: Individual results of the 10 participants in Experiment 2 in Chapter 4 for signals centered at 1500 Hz. Level difference at equal loudness between noise and a pure tone as a function of the bandwidth of the noise at a reference level of 30 dB SPL (black filled symbols), 50 dB SPL (gray filled symbols) and 70 dB SPL (open symbols). Error bars indicate intraindividual standard deviations. In each panel vertical lines show the critical bandwidth at a frequency of 1500 Hz in ERB$_N$ (dotted line) and Bark (dash-dotted line). To increase readability data are slightly shifted against each other.

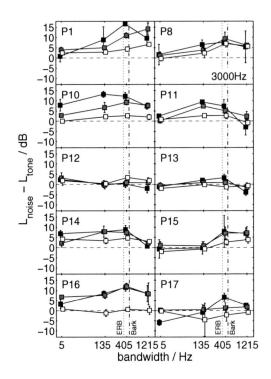

Figure 8.5: Individual results of the 10 participants in Experiment 2 in Chapter 4 for signals centered at 3000 Hz. Level difference at equal loudness between noise and a pure tone as a function of the bandwidth of the noise at a reference level of 30 dB SPL (black filled symbols), 50 dB SPL (gray filled symbols) and 70 dB SPL (open symbols). Error bars indicate intraindividual standard deviations. In each panel vertical lines show the critical bandwidth at a frequency of 3000 Hz in ERB_N (dotted line) and Bark (dash-dotted line). To increase readability data are slightly shifted against each other.

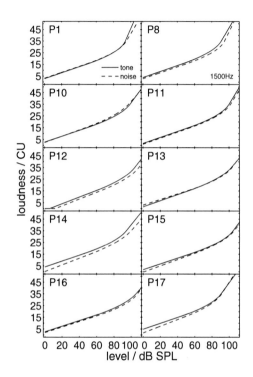

Figure 8.6: Mean loudness functions of the 10 participants in Experiment 3 in Chapter 4 for pure tones (solid lines) and running noise signals (dashed lines) centered at 1500 Hz.

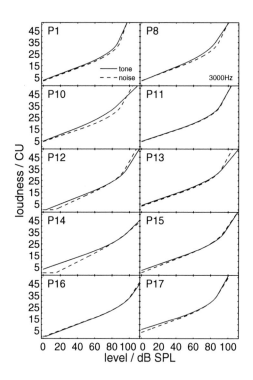

Figure 8.7: Mean loudness functions of the 10 participants in Experiment 3 in Chapter 4 for pure tones (solid lines) and running noise signals (dashed lines) centered at 3000 Hz.

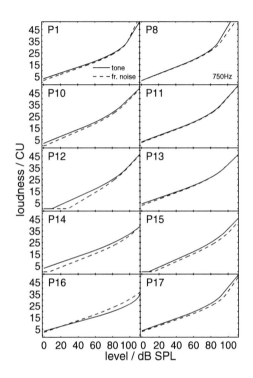

Figure 8.8: Mean loudness functions of the 10 participants in Experiment 3 in Chapter 4 for pure tones (solid lines) and frozen noise signals (dashed lines) centered at 750 Hz.

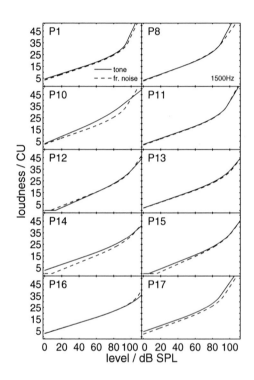

Figure 8.9: Mean loudness functions of the 10 participants in Experiment 3 in Chapter 4 for pure tones (solid lines) and frozen noise signals (dashed lines) centered at 1500 Hz.

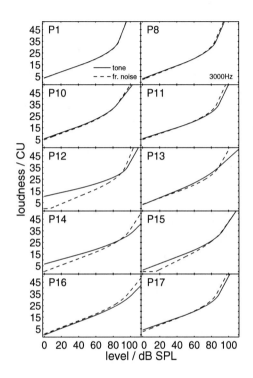

Figure 8.10: Mean loudness functions of the 10 participants in Experiment 3 in Chapter 4 for pure tones (solid lines) and frozen noise signals (dashed lines) centered at 3000 Hz.

8.3 Loudness functions and audiograms of the hearing-impaired participants in Chapter 5

Each panel of Fig. 8.11 shows the audiogram for one of the nine hearing-impaired participants in the experiment in Chapter 5 of the ear selected for the measurements (black symbols) and the other ear (gray symbols). The markers indicate the ears of the participant (crosses, left ear; circles, rigth ear). All participants showed a flat hearing loss between 35 and 75 dB HL in the investigated frequency range which is indicated by the gray area in each panel.

Figure 8.12 shows loudness functions for Gaussian noise stimuli geometrically centered at 1.5 kHz with a bandwidth of 135 Hz. Each panel shows the mean loudness function of the three repetitions for one of the nine hearing-impaired participants in Chapter 5 (solid black lines). For comparison, the averaged loudness function of the 10 normal-hearing participants in Experiment 3 in Chapter 4 for the Gaussian noise stimuli geometrically centered at 1.5 kHz with a bandwidth of 135 Hz from the running noise condition (i.e., the same type of stimuli) is included in each panel (dashed gray lines).

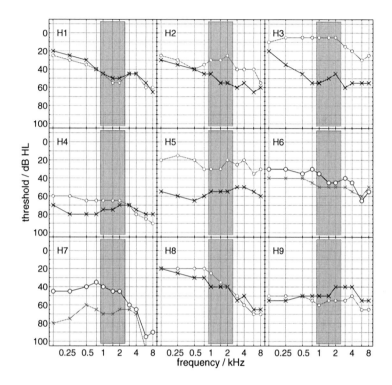

Figure 8.11: Audiograms of the nine hearing-impaired participants
in the experiment in Chapter 5 of the ear selected for
the measurements (black symbols) and the other ear
(gray symbols). Markers indicate the ear (crosses, left
ear; circles, rigth ear). The gray area indicates the
investigated frequency range.

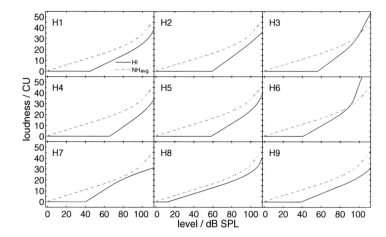

Figure 8.12: Mean loudness functions for Gaussian noise stimuli geometrically centered at 1.5 kHz with a bandwidth of 135 Hz of the nine hearing-impaired participants in the experiment in Chapter 5 (solid black lines). For comparison the averaged loudness function of the 10 normal-hearing listeners in Experiment 3 in Chapter 4 for the same type of stimuli is included in each panel (dashed gray lines).

Bibliography

ANSI S3.4 (**2007**). "Procedure for the computation of loudness of steady sounds", ANSI S3.4–2007.

Anweiler, A.K. and Verhey, J.L. (**2006**). "Spectral loudness summation for short and long signals as a function of level", J. Acoust. Soc. Am. **119**, 2919–2928.

Baker, R.J. and Rosen, S. (**2002**). "Auditory filter nonlinearity in mild/moderate hearing impairment", J. Acoust. Soc. Am. **111**, 1330–1339.

Bauch, H. (**1956**). "Die Bedeutung der Frequenzgruppe für die Lautheit von Klängen (The role of critical bands for the loudness of sounds)", Acustica **6**, 41–45.

Brand, T. and Hohmann, V. (**2001**). "Effect of hearing loss, centre frequency and bandwidth on the shape of loudness functions in categorical loudness scaling", Audiology **40**, 92–103.

Brand, T. and Hohmann, V. (**2002**). "An adaptive procedure for categroical loudness scaling", J. Acoust. Soc. Am. **112**, 1597–1604.

Buchholz, J.M., Caminade, S., Strelcyk, O. and Dau, T. (**2010**). "Estimating individual listeners' auditory-filter bandwidth in simultaneous and non-simultaneous masking", Proceedings of 20th International Congress on Acoustics, ICA 2010, Sydney, Australia, 1–7.

Buschermöhle, M., Verhey, J.L., Feudel, U. and Freund, J.A. (**2007**). "The role of the auditory periphery in comodulation detection difference and comodulation masking release", Biol. Cybern. **97**, 397–411.

Buus, S., Florentine, M. and Poulsen, T. (**1997**). "Temporal integration of loudness, loudness discrimination, and the form of the loudness function", J. Acoust. Soc. Am. **101**, 669–680.

Chalupper, S. and Fastl, H. (**2002**). "Dynamic loudness model (DLM) for normal and hearing-impaired listeners", Acta Acust. united Ac. **88**, 378–386.

Dau, T., Verhey, J.L. and Kohlrausch, A. (**1999**). "Intrinsic envelope fluctuations and modulation-detection thresholds for narrow-band noise carriers", J. Acoust. Soc. Am. **106**, 2752–2760.

DIN ISO 16832 (**2007**). "Acoustics – Loudness scaling by means of categories (ISO 16832:2006)".

DIN 45631/A1 (**2010**). "Calculation of loudness level and loudness from the sound spectrum - Zwicker method - Amendment 1: Calculation of the loudness of time-variant sound", DIN 45631/A1:2010–03.

Epstein, M. and Marozeau, J. (**2010**). "Loudness and intenstity coding", *The Oxford Handbook of Auditory Science – Hearing*, Volume 3, Edited by C.J. Plack, (Oxford University Press: Oxford, ISBN 978–0–19–923355–7) , 45–69.

Ernst, S.M.A., Rennies, J., Kollmeier, B. and Verhey, J.L. (**2010**). "Suppression and comodulation masking release in normal-hearing and hearing-impaired listeners", J. Acoust. Soc. Am. **128**, 300–309.

Ewert, S.E. (**2013**). "AFC - A modular framework for running psychoacoustic experiments and computational perception models", AIA-DAGA 2013, Merano, Italy. (Dega e.V., Berlin, Germany), 1326–1329.

Fastl, H. and Zwicker, E. (**2007**). "Psychoacoustics – Facts and Models", 3rd edition, Springer Berlin Heidelberg, ISBN 978-3-540-68888-4.

Fletcher, H. and Munson, W. (**1937**). "Relation Between Loudness and Masking", J. Acoust. Soc. Am. **9**, 1–10.

Florentine, M., Buus, S. and Bonding, P. (**1978**). "Loudness of complexs sounds as a function of the standard stimulus and the number of components", J. Acoust. Soc. Am. **64**, 1036–1040.

Florentine, M. (**2011**). "Loudness", in *Springer Handbook of Auditory Research – Loudness*, Volume 37, edited by M. Florentine, A.N. Popper and R.R. Fay, (Springer New York Dordrecht Heidelberg London, ISBN 978-1-4419-6711-4), 1–15.

Glasberg, B.R., Moore, B.C.J. and Bacon, S.P. (**1987**). "Gap detection and masking in hearing-impaired and normal-hearing subjects", J. Acoust. Soc. Am. **81**, 1546–1556.

Glasberg, B.R. and Moore, B.C.J. (**2002**). "A model of loudness applicable to time-varying sounds", J. Audio Eng. Soc. **50**, 331–341.

Grimm, G., Hohmann, V. and Verhey, J.L. (**2002**). "Loudness of fluctuating sounds", Acta Acust. united Ac. **88**, 359–368.

Hansen, H., Verhey, J.L. and Weber, R. (**2011**). "The Magnitude of Tonal Content. A Review", Acta Acust. united Ac. **97**, 355–363.

Harris, D.M. (**1979**). "Action potential suppression, tuning curves and thresholds: Comparison with single fiber data", Hearing Res. **1**, 133–154.

He, N., Dubno, J.R. and Mills, J. (**1998**). "Frequency and intensity discrimination measured in a maximum-likelihood procedure from young and aged normal-hearing subjects", J. Acoust. Soc. Am. **103**, 553-565.

ISO 226 (**2003**). "Acoustics – Normal equal-loudness-level contours", ISO 226–2003.

ISO 532 (**1975**). "Acoustics – Methods for calculating loudness level", ISO 532–1975.

Keller, J.N. (**2006**). "Loudness discomfort levels: A retrospective study comparing data from Pascoe (1988) and Washington University School of Medicine", *Independent Studies and Capstones*. Paper 83. Program in Audiology and Communication Sciences, Washington University School of Medicine. http://digitalcommons.wustl.edu/pacs_capstones/83

Kujawa, S.G. and Liberman, M.C. (**2009**). "Adding insult to injury: cochlear nerve degeneration after temporary noise-induced hearing loss", J. Neurosci. **29**, 14077–14085.

Kumagai, M., Suzuki, Y. and Sone, T. (**1984**). "A study on the time constant for an impulse sound level meter (A study on the loudness of impact sound. V)", J. Acoust. Soc. Jpn. **5**, 31–36.

Launer, S. (**1995**). "Loudness perception in listeners with sensorineural hearing impairment", Dissertation am Fachbereich Physik, Universität Oldenburg. http://medi.uni-oldenburg.de/download/docs/diss/Launer_1995_LoudnessPerception/

Lawson, J.L. and Uhlenbeck, G.E. (**1950**). "Threshold Signals", Vol. 24 of Radiation Laboratory Series (McGraw-Hill, New York), Chapter 3.8, 56–63.

Leek, M.R. and Summers, V. (**2001**). "Pitch strength and pitch dominance of iterated rippled noises in hearing-impaired listeners", J. Acoust. Soc. Am. **109**, 2954–2944.

Lecluyse, W. and Meddis, R. (**2009**). "A simple single-interval adaptive procedure for estimating thresholds in normal and impaired listeners.", J. Acoust. Soc. Am. **126**, 2570–2579.

Levitt, H. (**1971**). "Transformed Up-Down Methods in Psychoacoustics", J. Acoust. Soc. Am. **49**, 467–477.

Liberman, M.C. and Kiang, N.Y.S. (**1978**). "Acoustic trauma in cats - cochlear pathology and auditory-nerve activity", Acta Oto-Laryngologica **358**, 5–63.

Lutfi, R.A. and Patterson, R.D. (**1984**). "On the growth of masking asymmetry with stimulus intensity", J. Acoust. Soc. Am. **76**, 739–745.

Moore, B.C.J. and Glasberg, B.R. (**1981**). "Auditory filter shapes derived in simultaneous and forward masking", J. Acoust. Soc. Am. **70**, 1003–1014.

Moore, B.C.J., Wojtczak, M. and Vickers, D.A. (**1996**). "Effect of loudness recruitment on the perception of amplitude modulation", J. Acoust. Soc. Am. **100**, 481–489.

Moore, B.C.J., Glasberg, B.R. and Baer, T. (**1997**). "A model for the prediction of thresholds, loudness and partial loudness", J. Audio Eng. Soc. **3**, 289–311.

Moore, B.C.J. and Glasberg, B.R. (**1997**). "A model of loudness perception applied to cochlear hearing loss", Auditory Neuroscience **82**, 335–345.

Moore, B.C.J., Vickers, D.A., Baer, T. and Launer, S. (**1999**). "Factors affecting the loudness of modulated sounds", J. Acoust. Soc. Am. **105**, 2757–2772.

Moore, B.C.J. (**2003**). "An introduction to the psychology of hearing", 5th edition, Academic Press, San Diego, ISBN 978-0-12-505628-1.

Moore, B.C.J. and Glasberg, B.R. (**2007**). "Modeling binaural loudness", J. Acoust. Soc. Am. **121**, 1604–1612.

Morita, T., Naito, Y., Nagamine, T., Fujiki, N., Shibasaki, H. and Ito, J. (**2003**). "Enhanced activation of the auditory cortex in patients with inner-ear hearing impairment: A magnetoencephalographic study", Clin. Neurophysiol. **114**, 851–859.

Nitschmann, M., Verhey, J.L. and Kollmeier, B. (**2010**). "Monaural and binaural frequency selectivity in hearing-impaired subjects", Int. J. Audiol. **49**, 357–367.

Norena, A.J., Micheyl, C., Chery-Croze, S. and Collet, L. (**2002**). "Psychoacoustic characterization of the tinnitus spectrum: implications for the underlying mechanisms of tinnitus", Audiol. Neurootol. **7**, 358–369.

Oxenham, A.J., Moore, B.C.J. and Vickers, D.A. (**1997**). "Short-term temporal integration: Evidence for the influence of peripheral compression", J. Acoust. Soc. Am. **101**, 3676–3687.

Pedersen, O.J., Lyregaard, P.E. and Poulsen, T. (**1977**). "The round robin test on impulsive noise", Rep. No. **22**, The Acoustics Laboratory, Technical Universtiy of Denmark, Lyngby, Denmark, 1–180.

Poulsen, T. (**1981**). "Loudness of tone pulses in a free field", J. Acoust. Soc. Am. **69**, 1786–1790.

Rennies, J., Verhey, J.L., Chalupper, S. and Fastl, H. (**2009**). "Modeling temporal effects of spectral loudness summation", Acta Acust. united Ac. **95**, 1112–1122.

Rennies, J., Verhey, J.L. and Fastl, H. (**2010**). "Comparison of loudness models for time-varying sounds", Acta Acust. united Ac. **96**, 383–396.

Rennies, J., Holube, I., and Verhey, J.L. (**2013**). "Loudness of speech and speech-like signals", Acta Acust. united Ac. **99**, 268–282.

Schaette, R. and Kempter, R. (**2009**). "Predicting tinnitus pitch from patients' audiograms with a computational model for the development of neuronal hyperactivity.", J. Neurophysiol. **101**, 3042–3052.

Schaette, R. and McAlpine, D. (**2011**). "Tinnitus with a normal audiogram: physiological evidence for hidden hearing loss and computational model", J. Neurosci. **31**, 13452–13457.

Scharf, B. (**1978**). "Loudness", in *Handbook of Auditory Perception Vol. IV, Hearing*, edited by E.C. Caterette and M.P. Friedman, (Academic, New York, ISBN 0–12–161904–4), 187–242.

Schmiedt, R.A. (**1982**). "Boundaries of two-tone rate suppression of cochlear-nerve activity", Hearing Res. **7**, 335–351.

Sellick, P.M. and Russel, I.J. (**1979**). "Two-tone suppression in cochlear hair cells", Hearing Res. **1**, 227–236.

Shannon, R.V. (**1976**). "Two-tone unmasking and suppression in a forward-masking situation", J. Acoust. Soc. Am. **59**, 1460–1470.

Soeta, Y., Nakagawa, S. and Tonoike, M. (**2005**). "Auditory evoked magnetic fields in relation to bandwidth variations of bandpass noise", Hearing Res. **202**, 47–54.

Soeta, Y. and Nakagawa, S. (**2006**). "Complex tone processing and critical band in the human auditory cortex", Hearing Res. **222**, 125–132.

Sottek, R. (**2013**). "Progresses in standardizing loudness of time-variant sounds", AIA-DAGA 2013, Merano, Italy (Dega e.V., Berlin, Germany), ISBN: 978-3-939296-05-8, 1125–1128.

Sutter, M.L. and Loftus, W.C. (**2003**). "Exitatory and inhibitory intensity tuning in auditory cortex: Evidence for multiple inhibitory mechanisms", J. Neurophysiol. **90**, 2629–2647.

Valente, D.L., Joshi, S.N. and Jesteadt, W. (**2011**). "Temporal integration of loudness measured using categorical loudness scaling and matching procedures", J. Acoust. Soc. Am. **130**, EL32–EL37.

Verhey, J.L. and Kollmeier, B. (**2002**). "Spectral loudness summation as a function of duration", J. Acoust. Soc. Am. **111**, 1349–1358.

Verhey, J.L., Anweiler, A.K. and Hohmann, V. (**2006**). "Spectral loudness summation as a function of duration for hearing-impaired listeners", Int. J. Audiol. **45**, 287–294.

Verhey, J.L. and Uhlemann, M. (**2008**). "Spectral loudness summation for sequences of short noise bursts", J. Acoust. Soc. Am. **123**, 925–934.

Verhey, J.L. (**2010**). "Temporal resolution and temporal integration", *The Oxford Handbook of Auditory Science – Hearing*, Volume 3, Edited by C.J. Plack, (Oxford University Press: Oxford, ISBN 978-0-19-923355-7) , 105–121.

Viemeister, N.F. and Bacon, S.P. (**1988**). "Intensity discrimination, increment detection, and magnitude estimation for 1-kHz tones", J. Acoust. Soc. Am. **84**, 172–178.

Viemeister, N.F. and Wakefield, G.H. (**1991**). "Temporal integration and multiple looks", J. Acoust. Soc. Am. **90**, 858–865.

Weisz, N., Hartmann, T., Dohrmann, K. and Schlee, W. Norena, A.J. (**2006**). "High-frequency tinnitus without hearing loss does not mean absence of deafferentation", Hear. Res. **222**, 108–114.

Yates, G.K., Winter, I.M., and Robertson, D. (**1990**). "Basilar membrane nonlinearity determines auditory nerve rate-intensity functions and cochlear dynamic range", Hear. Res. **45**, 203–219.

Zhang, C. and Zeng, F. (**1997**). "Loudness of dynamic stimuli in acoustic and electric hearing", J. Acoust. Soc. Am. **102**, 2925–2945.

Zwicker, E., Flottorp, G. and Stevens, S.S. (**1957**). "Critical band width in loudness summation", J. Acoust. Soc. Am. **29**, 548–557.

Zwicker, E. (**1958**). "Über psychologische und methodische Grundlagen der Lautheit (On psychological and methodical principles of loudness)", Acustica **8**, 237–258.

Zwicker, E. (**1961**). "Subdivision of the Audible Frequency Range into Critical Bands (Frequenzgruppen)", J. Acoust. Soc. Am. **33**, 248.

Zwicker, E. and Scharf, B. (**1965**). "A model of loudness summation", Psychological review **72**, 3–26.

Zwicker, E. (**1974**). "Loudness and excitation patterns of strongly frequency modulated tones", in Sensation and Measurement, papers in honor of S.S. Stevens, edited by H.R. Moskowitz, B. Scharf, and J.C. Stevens (D. Reidel, Dordrecht/Boston), 325–335.

List of Figures

List of Tables

Danksagungen

Besonders herzlich möchte ich Prof. Dr. Jesko L. Verhey für die umfangreiche Betreuung meiner Arbeit danken. Während der gesamten Zeit meiner Arbeit konnte ich mich auf seine Hilfe stets verlassen.

Bedanken möchte ich mich bei Prof. Dr. Torsten Dau für die Begutachtung meiner Arbeit und den weiteren Mitgliedern der Prüfungskommission Prof. Dr. Jan Wiersig als Komissionsvorsitzenden sowie Prof. Dr. Oliver Speck und PD Dr. Peter Heil.

Ein großes Dankeschön geht an Jan Rennies, Bastian Epp und Roland Schaette für die angenehme Zusammenarbeit. Roland Schaette danke ich auch für die gute Zeit in London während der Messungen für die in Kapitel 6 vorgestellte Studie. Brian C. J. Moore und einigen anonymen Reviewern möchte ich für ihre kritische Auseinandersetzung mit den eingereichten Manuskripten der veröffentlichten Teile dieser Arbeit danken.

Auch der Abteilung für Experimentelle Audiologie möchte ich danken, die mir in ihrer lockeren Art immer zur Seite stand und es mir ermöglicht hat Auszüge meiner Arbeit auf verschiedenen internationalen Tagungen vorzustellen. Insbesondere gilt mein Dank hier Katrin Mentzel, die mir eine große Hilfe bei der Durchführung der in Kapitel 5 beschriebenen Experimente war.

Besonders gedankt sei auch allen Versuchspersonen für die bereitwillige Teilnahme an den oft konzentrationsintensiven Versuchsreihen.

Zu guter Letzt gilt mein Dank natürlich all den Menschen, die mich durch die Zeit meiner Promotion begleitet und diese sehr bereichert haben.